Elliptic operators, topology and asymptotic methods

Second Edition

CHAPMAN & HALL/CRC
Research Notes in Mathematics Series

Main Editors

H. Brezis, *Université de Paris*
R.G. Douglas, *Texas A&M University*
A. Jeffrey, *University of Newcastle upon Tyne (Founding Editor)*

Editorial Board

H. Amann, *University of Zürich*
R. Aris, *University of Minnesota*
G.I. Barenblatt, *University of Cambridge*
H. Begehr, *Freie Universität Berlin*
P. Bullen, *University of British Columbia*
R.J. Elliott, *University of Alberta*
R.P. Gilbert, *University of Delaware*
D. Jerison, *Massachusetts Institute of Technology*
B. Lawson, *State University of New York at Stony Brook*

B. Moodie, *University of Alberta*
S. Mori, *Kyoto University*
L.E. Payne, *Cornell University*
D.B. Pearson, *University of Hull*
I. Raeburn, *University of Newcastle, Australia*
G.F. Roach, *University of Strathclyde*
I. Stakgold, *University of Delaware*
W.A. Strauss, *Brown University*
J. van der Hoek, *University of Adelaide*

Submission of proposals for consideration

Suggestions for publication, in the form of outlines and representative samples, are invited by the Editorial Board for assessment. Intending authors should approach one of the main editors or another member of the Editorial Board, citing the relevant AMS subject classifications. Alternatively, outlines may be sent directly to the publisher's offices. Refereeing is by members of the board and other mathematical authorities in the topic concerned, throughout the world.

Preparation of accepted manuscripts

On acceptance of a proposal, the publisher will supply full instructions for the preparation of manuscripts in a form suitable for direct photo-lithographic reproduction. Specially printed grid sheets can be provided. Word processor output, subject to the publisher's approval, is also acceptable.

Illustrations should be prepared by the authors, ready for direct reproduction without further improvement. The use of hand-drawn symbols should be avoided wherever possible, in order to obtain maximum clarity of the text.

The publisher will be pleased to give guidance necessary during the preparation of a typescript and will be happy to answer any queries.

Important note

In order to avoid later retyping, intending authors are strongly urged not to begin final preparation of a typescript before receiving the publisher's guidelines. In this way we hope to preserve the uniform appearance of the series.

CRC Press UK
Chapman & Hall/CRC Statistics and Mathematics
Pocock House
235 Southwark Bridge Road
London SE1 6LY
Tel: 020 7450 7335

Titles in this series. A full list is available from the publisher on request.

251 Stability of stochastic differential equations with respect to semimartingales
   X Mao
252 Fixed point theory and applications
   J Baillon and M Théra
253 Nonlinear hyperbolic equations and field theory
   M K V Murthy and S Spagnolo
254 Ordinary and partial differential equations. Volume III
   B D Sleeman and R J Jarvis
255 Harmonic maps into homogeneous spaces
   M Black
256 Boundary value and initial value problems in complex analysis: studies in complex analysis and its applications to PDEs 1
   R Kühnau and W Tutschke
257 Geometric function theory and applications of complex analysis in mechanics: studies in complex analysis and its applications to PDEs 2
   R Kühnau and W Tutschke
258 The development of statistics: recent contributions from China
   X R Chen, K T Fang and C C Yang
259 Multiplication of distributions and applications to partial differential equations
   M Oberguggenberger
260 Numerical analysis 1991
   D F Griffiths and G A Watson
261 Schur's algorithm and several applications
   M Bakonyi and T Constantinescu
262 Partial differential equations with complex analysis
   H Begehr and A Jeffrey
263 Partial differential equations with real analysis
   H Begehr and A Jeffrey
264 Solvability and bifurcations of nonlinear equations
   P Drábek
265 Orientational averaging in mechanics of solids
   A Lagzdins, V Tamuzs, G Teters and A Kregers
266 Progress in partial differential equations: elliptic and parabolic problems
   C Bandle, J Bemelmans, M Chipot, M Grüter and J Saint Jean Paulin
267 Progress in partial differential equations: calculus of variations, applications
   C Bandle, J Bemelmans, M Chipot, M Grüter and J Saint Jean Paulin
268 Stochastic partial differential equations and applications
   G Da Prato and L Tubaro
269 Partial differential equations and related subjects
   M Miranda
270 Operator algebras and topology
   W B Arveson, A S Mishchenko, M Putinar, M A Rieffel and S Stratila
271 Operator algebras and operator theory
   W B Arveson, A S Mishchenko, M Putinar, M A Rieffel and S Stratila
272 Ordinary and delay differential equations
   J Wiener and J K Hale
273 Partial differential equations
   J Wiener and J K Hale
274 Mathematical topics in fluid mechanics
   J F Rodrigues and A Sequeira
275 Green functions for second order parabolic integro-differential problems
   M G Garroni and J F Menaldi
276 Riemann waves and their applications
   M W Kalinowski
277 Banach C(K)-modules and operators preserving disjointness
   Y A Abramovich, E L Arenson and A K Kitover
278 Limit algebras: an introduction to subalgebras of C^*-algebras
   S C Power
279 Abstract evolution equations, periodic problems and applications
   D Daners and P Koch Medina
280 Emerging applications in free boundary problems
   J Chadam and H Rasmussen
281 Free boundary problems involving solids
   J Chadam and H Rasmussen
282 Free boundary problems in fluid flow with applications
   J Chadam and H Rasmussen
283 Asymptotic problems in probability theory: stochastic models and diffusions on fractals
   K D Elworthy and N Ikeda
284 Asymptotic problems in probability theory: Wiener functionals and asymptotics
   K D Elworthy and N Ikeda
285 Dynamical systems
   R Bamon, R Labarca, J Lewowicz and J Palis
286 Models of hysteresis
   A Visintin
287 Moments in probability and approximation theory
   G A Anastassiou
288 Mathematical aspects of penetrative convection
   B Straughan
289 Ordinary and partial differential equations. Volume IV
   B D Sleeman and R J Jarvis
290 K-theory for real C^*-algebras
   H Schröder
291 Recent developments in theoretical fluid mechanics
   G P Galdi and J Necas
292 Propagation of a curved shock and nonlinear ray theory
   P Prasad
293 Non-classical elastic solids
   M Ciarletta and D Ieşan
294 Multigrid methods
   J Bramble
295 Entropy and partial differential equations
   W A Day
296 Progress in partial differential equations: the Metz surveys 2
   M Chipot
297 Nonstandard methods in the calculus of variations
   C Tuckey
298 Barrelledness, Baire-like- and (LF)-spaces
   M Kunzinger
299 Nonlinear partial differential equations and their applications. Collège de France Seminar. Volume XI
   H Brezis and J L Lions

300 Introduction to operator theory
  T Yoshino
301 Generalized fractional calculus and applications
  V Kiryakova
302 Nonlinear partial differential equations and their applications. Collège de France Seminar Volume XII
  H Brezis and J L Lions
303 Numerical analysis 1993
  D F Griffiths and G A Watson
304 Topics in abstract differential equations
  S Zaidman
305 Complex analysis and its applications
  C C Yang, G C Wen, K Y Li and Y M Chiang
306 Computational methods for fluid-structure interaction
  J M Crolet and R Ohayon
307 Random geometrically graph directed self-similar multifractals
  L Olsen
308 Progress in theoretical and computational fluid mechanics
  G P Galdi, J Málek and J Necas
309 Variational methods in Lorentzian geometry
  A Masiello
310 Stochastic analysis on infinite dimensional spaces
  H Kunita and H-H Kuo
311 Representations of Lie groups and quantum groups
  V Baldoni and M Picardello
312 Common zeros of polynomials in several variables and higher dimensional quadrature
  Y Xu
313 Extending modules
  N V Dung, D van Huynh, P F Smith and R Wisbauer
314 Progress in partial differential equations: the Metz surveys 3
  M Chipot, J Saint Jean Paulin and I Shafrir
315 Refined large deviation limit theorems
  V Vinogradov
316 Topological vector spaces, algebras and related areas
  A Lau and I Tweddle
317 Integral methods in science and engineering
  C Constanda
318 A method for computing unsteady flows in porous media
  R Raghavan and E Ozkan
319 Asymptotic theories for plates and shells
  R P Gilbert and K Hackl
320 Nonlinear variational problems and partial differential equations
  A Marino and M K V Murthy
321 Topics in abstract differential equations II
  S Zaidman
322 Diffraction by wedges
  B Budaev
323 Free boundary problems: theory and applications
  J I Diaz, M A Herrero, A Liñan and J L Vazquez
324 Recent developments in evolution equations
  A C McBride and G F Roach
325 Elliptic and parabolic problems: Pont-à-Mousson 1994
  C Bandle, J Bemelmans, M Chipot, J Saint Jean Paulin and I Shafrir

326 Calculus of variations, applications and computations: Pont-à-Mousson 1994
  C Bandle, J Bemelmans, M Chipot, J Saint Jean Paulin and I Shafrir
327 Conjugate gradient type methods for ill-posed problems
  M Hanke
328 A survey of preconditioned iterative methods
  A M Bruaset
329 A generalized Taylor's formula for functions of several variables and certain of its applications
  J-A Riestra
330 Semigroups of operators and spectral theory
  S Kantorovitz
331 Boundary-field equation methods for a class of nonlinear problems
  G N Gatica and G C Hsiao
332 Metrizable barrelled spaces
  J C Ferrando, M López Pellicer and L M Sánchez Ruiz
333 Real and complex singularities
  W L Marar
334 Hyperbolic sets, shadowing and persistence for noninvertible mappings in Banach spaces
  B Lani-Wayda
335 Nonlinear dynamics and pattern formation in the natural environment
  A Doelman and A van Harten
336 Developments in nonstandard mathematics
  N J Cutland, V Neves, F Oliveira and J Sousa-Pinto
337 Topological circle planes and topological quadrangles
  A E Schroth
338 Graph dynamics
  E Prisner
339 Localization and sheaves: a relative point of view
  P Jara, A Verschoren and C Vidal
340 Mathematical problems in semiconductor physics
  P Marcati, P A Markowich and R Natalini
341 Surveying a dynamical system: a study of the Gray–Scott reaction in a two-phase reactor
  K Alhumaizi and R Aris
342 Solution sets of differential equations in abstract spaces
  R Dragoni, J W Macki, P Nistri and P Zecca
343 Nonlinear partial differential equations
  A Benkirane and J-P Gossez
344 Numerical analysis 1995
  D F Griffiths and G A Watson
345 Progress in partial differential equations: the Metz surveys 4
  M Chipot and I Shafrir
346 Rings and radicals
  B J Gardner, Liu Shaoxue and R Wiegandt
347 Complex analysis, harmonic analysis and applications
  R Deville, J Esterle, V Petkov, A Sebbar and A Yger
348 The theory of quantaloids
  K I Rosenthal
349 General theory of partial differential equations and microlocal analysis
  Qi Min-you and L Rodino

350 Progress in elliptic and parabolic partial differential equations
   A Alvino, P Buonocore, V Ferone, E Giarrusso, S Matarasso, R Toscano and G Trombetti
351 Integral representations for spatial models of mathematical physics
   V V Kravchenko and M V Shapiro
352 Dynamics of nonlinear waves in dissipative systems: reduction, bifurcation and stability
   G Dangelmayr, B Fiedler, K Kirchgässner and A Mielke
353 Singularities of solutions of second order quasilinear equations
   L Véron
354 Mathematical theory in fluid mechanics
   G P Galdi, J Málek and J Necas
355 Eigenfunction expansions, operator algebras and symmetric spaces
   R M Kauffman
356 Lectures on bifurcations, dynamics and symmetry
   M Field
357 Noncoercive variational problems and related results
   D Goeleven
358 Generalised optimal stopping problems and financial markets
   D Wong
359 Topics in pseudo-differential operators
   S Zaidman
360 The Dirichlet problem for the Laplacian in bounded and unbounded domains
   C G Simader and H Sohr
361 Direct and inverse electromagnetic scattering
   A H Serbest and S R Cloude
362 International conference on dynamical systems
   F Ledrappier, J Lewowicz and S Newhouse
363 Free boundary problems, theory and applications
   M Niezgódka and P Strzelecki
364 Backward stochastic differential equations
   N El Karoui and L Mazliak
365 Topological and variational methods for nonlinear boundary value problems
   P Drábek
366 Complex analysis and geometry
   V Ancona, E Ballico, R M Mirò-Roig and A Silva
367 Integral expansions related to Mehler–Fock type transforms
   B N Mandal and N Mandal
368 Elliptic boundary value problems with indefinite weights: variational formulations of the principal eigenvalue and applications
   F Belgacem
369 Integral transforms, reproducing kernels and their applications
   S Saitoh
370 Ordinary and partial differential equations. Volume V
   P D Smith and R J Jarvis
371 Numerical methods in mechanics
   C Conca and G N Gatica
372 Generalized manifolds
   K-G Schlesinger
373 Independent axioms for Minkowski space-time
   J W Schutz
374 Integral methods in science and engineering Volume one: analytic methods
   C Constanda, J Saranen and S Seikkala
375 Integral methods in science and engineering Volume two: approximation methods
   C Constanda, J Saranen and S Seikkala
376 Inner product spaces and applications
   T M Rassias
377 Functional analysis with current applications in science, technology and industry
   M Brokate and A H Siddiqi
378 Classical and quantic periodic motions of multiply polarized spin-particles
   A Bahri
379 Analysis, numerics and applications of differential and integral equations
   M Bach, C Constanda, G C Hsiao, A-M Sändig and P Werner
380 Numerical analysis 1997
   D F Griffiths, D J Higham and G A Watson
381 Real analytic and algebraic singularities
   T Fukuda, T Fukui, S Izumiya and S Koike
382 Boundary value problems with equivalued surface and resistivity well-logging
   T Li, S Zheng, Y Tan and W Shen
383 Progress in partial differential equations Pont-à-Mousson 1997 Volume 1
   H Amann, C Bandle, M Chipot, F Conrad and I Shafrir
384 Progress in partial differential equations Pont-à-Mousson 1997 Volume 2
   H Amann, C Bandle, M Chipot, F Conrad and I Shafrir
385 The linear theory of Colombeau generalized functions
   M Nedeljkov, S Pilipovic and D Scarpalézos
386 Recent advances in differential equations
   H-H Dai and P L Sachdev
387 Progress in holomorphic dynamics
   H Kriete
388 Navier–Stokes equations: theory and numerical methods
   R Salvi
389 Strongly irreducible operators on Hilbert space
   C Jiang and Z Wang
390 Mathematical methods in scattering theory and biomedical technology
   G Dassios, D I Fotiadis, K Kiriaki and C V Massalas
391 Nonlinear partial differential equations and their applications. Collège de France Seminar Volume XIII
   D Cioranescu and J L Lions
392 Advanced topics in theoretical fluid mechanics
   J Málek, J Necas and M Rokyta
393 Topics in random polynomials
   K Farahmand
394 Dirac operators in analysis
   J Ryan and D Struppa
395 Elliptic operators, topology and asymptotic methods, Second Edition
   John Roe

John Roe
Pennsylvania State University

Elliptic operators, topology and asymptotic methods

Second Edition

CHAPMAN & HALL/CRC

Boca Raton   London   New York   Washington, D.C.

This book contains information obtained from authentic and highly regarded sources. Reprinted material is quoted with permission, and sources are indicated. A wide variety of references are listed. Reasonable efforts have been made to publish reliable data and information, but the author and the publisher cannot assume responsibility for the validity of all materials or for the consequences of their use.

Apart from any fair dealing for the purpose of research or private study, or criticism or review, as permitted under the UK Copyright Designs and Patents Act, 1988, this publication may not be reproduced, stored or transmitted, in any form or by any means, electronic or mechanical, including photocopying, microfilming, and recording, or by any information storage or retrieval system, without the prior permission in writing of the publishers, or in the case of reprographic reproduction only in accordance with the terms of the licenses issued by the Copyright Licensing Agency in the UK, or in accordance with the terms of the license issued by the appropriate Reproduction Rights Organization outside the UK.

The consent of CRC Press LLC does not extend to copying for general distribution, for promotion, for creating new works, or for resale. Specific permission must be obtained in writing from CRC Press LLC for such copying.

Direct all inquiries to CRC Press LLC, 2000 N.W. Corporate Blvd., Boca Raton, Florida 33431.

Trademark Notice: Product or corporate names may be trademarks or registered trademarks, and are used only for identification and explanation, without intent to infringe.

Visit the CRC Press Web site at www.crcpress.com

© Addison Wesley Longman Limited 1988, 1998
First Published 1988
Second Edition 1998
First CRC Reprint 2001
Originally published by Addison Wesley Longman

No claim to original U.S. Government works
International Standard Book Number 0-582-32502-1
Printed in the United States of America   3 4 5 6 7 8 9 0
Printed on acid-free paper

to Liane

Contents

Chapter 1. Resumé of Riemannian geometry	9
Connections	9
Riemannian geometry	12
Differential forms	17
Exercises	21
Chapter 2. Connections, curvature, and characteristic classes	23
Principal bundles and their connections	23
Characteristic classes	29
Genera	34
Notes	37
Exercises	37
Chapter 3. Clifford algebras and Dirac operators	41
Clifford bundles and Dirac operators	41
Clifford bundles and curvature	46
Examples of Clifford bundles	49
Notes	52
Exercises	53
Chapter 4. The Spin groups	55
The Clifford algebra as a superalgebra	55
Groups of invertibles in the Clifford algebra	57
Representation theory of the Clifford algebra	59
Spin structures on manifolds	62
Spin bundles and characteristic classes	64
The complex Spin group	67
Notes	68
Exercises	68
Chapter 5. Analytic properties of Dirac operators	71

Sobolev Spaces	71
Analysis of the Dirac operator	75
The functional calculus	82
Notes	84
Exercises	84

Chapter 6. Hodge theory   87
    Notes   91
    Exercises   92

Chapter 7. The heat and wave equations   95
    Existence and uniqueness theorems   95
    The asymptotic expansion for the heat kernel   99
    Finite propagation speed for the wave equation   104
    Notes   107
    Exercises   108

Chapter 8. Traces and eigenvalue asymptotics   109
    Eigenvalue growth   109
    Trace-class operators   110
    Weyl's asymptotic formula   114
    Notes   117
    Exercises   117

Chapter 9. Some non-compact manifolds   119
    The harmonic oscillator   119
    Witten's perturbation of the de Rham complex   123
    Functional calculus on open manifolds   127
    Notes   130
    Exercises   130

Chapter 10. The Lefschetz formula   13.
    Lefschetz numbers   13
    The fixed-point contributions   13
    Notes   14
    Exercises   14

Chapter 11. The index problem   14
    Gradings and Clifford bundles   14
    Graded Dirac operators   14

The heat equation and the index theorem	147
Notes	148
Exercises	149

Chapter 12. The Getzler calculus and the local index theorem — 151
- Filtered algebras and symbols — 151
- Getzler symbols — 154
- The Getzler symbol of the heat kernel — 157
- The exact solution — 162
- The index theorem — 164
- Notes — 165
- Exercises — 166

Chapter 13. Applications of the index theorem — 169
- The spinor Dirac operator — 169
- The signature theorem — 172
- The Hirzebruch-Riemann-Roch theorem — 175
- Local index theory — 177
- Notes — 179
- Exercises — 180

Chapter 14. Witten's approach to Morse theory — 183
- The Morse inequalities — 183
- Morse functions — 185
- The contribution from the critical points — 189
- Notes — 192

Chapter 15. Atiyah's Γ-index theorem — 193
- An algebra of smoothing operators — 193
- Renormalized dimensions and the index theorem — 198
- Notes — 201

References — 203

Introduction

This book, loosely based on a course of lectures delivered at Oxford in 1987, is intended as an introduction to the circle of ideas surrounding the heat equation proof of the Atiyah-Singer index theorem. Among the topics discussed are Hodge theory; the asymptotic expansion for the heat kernel; Weyl's theorem on the distribution of the eigenvalues of the Laplacian; the index theorem for Dirac-type operators, using Getzler's direct method; Witten's analytic approach to the Morse inequalities; and the L^2-index theorem of Atiyah for Galois coverings. As background one needs an acquaintance with differential geometry (connections, metrics, curvature, exterior derivative, de Rham cohomology) and functional analysis (elementary theory of Banach, Hilbert, and occasionally Frechet spaces; the spectral theorem for compact self-adjoint operators). Almost all the PDE theory needed is developed in the text. Occasionally we quote results from representation theory or algebraic topology, but the reader unfamiliar with these should be able to skip over them without loss of continuity. Some exercises are provided, which introduce a number of topics not treated in the text.

In the ten years since the first edition of this book appeared there have been a number of much more comprehensive treatments of this material, among which I have found the works by Berline-Getzler-Vergne [12] and Lawson-Michelsohn [47] especially useful. The book is now organized as a three-course meal: four chapters of geometry (1–4), and five chapters of analysis (5–9), culminate in four chapters of topology (10–13) in which the preceding results are brought together to prove first the Lefschetz formula and then the full index theorem. The final two chapters (14–15) are a dessert.

Once again I am grateful to everyone who has shared insights, comments, or suggestions about this book, both before and after the publication of the first edition. I am grateful also to Ivan Smith and Jon Woolf who undertook the task of producing a TEX file of the first edition; I trust that enough of their work survives in this second edition that they feel their efforts to have been worthwhile!

JOHN ROE

OXFORD AND PENN STATE, JULY 1998

CHAPTER 1

Resumé of Riemannian geometry

In this chapter we give a rapid tour through some of the basic ideas of Riemannian geometry. For the most part we will omit proofs if they can be found in the beautiful summary by Milnor [54, Part II].

Connections

Let M be a smooth manifold, and let V be a vector bundle on M. By $C^\infty(V)$ we denote the space of smooth sections of V. In particular, TM will denote the tangent bundle of M, and $C^\infty(TM)$ will denote the space of smooth sections of TM, also known as *vector fields* on M.

DEFINITION 1.1 A *connection* on V is a linear map

$$\nabla \colon C^\infty(TM) \otimes C^\infty(V) \to C^\infty(V)$$

assigning to a vector field X and a section Y of V a new vector field $\nabla_X Y$ such that, for any smooth function f on M,

(i) $\nabla_{fX} Y = f \nabla_X Y$

(ii) $\nabla_X(fY) = f \nabla_X Y + (X.f)Y$, where $X.f$ denotes the Lie derivative of f along X.

Condition (i) may be expressed more technically by saying that $X \mapsto \nabla_X Y$ is a homomorphism of modules over $C^\infty(M)$. It shows that the value of $\nabla_X Y$ at a point depends only on the value of X at p. This allows us to regard ∇ as a map from $C^\infty(V)$ to $\Omega^1(V) := C^\infty(T^*M \otimes V)$, the space of V-valued one-forms, and it is often useful to think of a connection in this way. Condition (ii) is modeled on the product rule for differentiation. In fact, if V is trivialized (as can always be done locally), then vector fields are identified with $\mathbb{R}^k$-valued functions and an example of a connection is given by

$$\nabla_X = \text{Lie derivative along } X.$$

Thus many connections exist locally. Moreover, the *difference* of two connections is local both in X and Y, so is an $\text{End}(V)$-valued one-form; in other words, the space of connections is an affine space modeled on the vector space $\Omega^1(\text{End}(V))$. Now using a partition of unity it is easy to see that local connections can be patched together to give a global connection. So, every vector bundle (over a paracompact manifold) admits connections.

REMARK 1.2 Suppose that V is trivialized over a coordinate patch of M, with local coordinates $x^1, \ldots, x^n$. Then from our discussion above, for any connection ∇,

$$\nabla_i = \partial/\partial x^i + \Gamma_i$$

where Γ_i is a smooth section of $\text{End}(V)$. (We abbreviate ∇_{∂_i} to ∇_i.) The functions Γ_i are called the *Christoffel symbols* and they determine the connection completely. Notice that they depend both on the local coordinate system for M and on the chosen trivialization of V.

REMARK 1.3 An alternative definition of connections starts from the idea of *parallel transport*. Let γ be a smooth curve in M. The differential equation $\nabla_{\dot\gamma} Y = 0$ is a first order ordinary differential equation on sections Y of V along γ, which has a unique solution for a given initial value; one says that Y is *parallel* along γ, or that it has been obtained from its initial value by *parallel transport*. The connection determines the notion of parallel transport, and conversely parallel transport determines the connection ∇, as we shall see later.

DEFINITION 1.4 The *curvature operator* K of a connection ∇ on V is defined as follows: if X, Y are vector fields and Z is a section of V, then

$$K(X,Y)Z = \nabla_X \nabla_Y Z - \nabla_Y \nabla_X Z - \nabla_{[X,Y]} Z.$$

It would appear from its definition that K is a differential operator. However the following well-known calculation (proof in [54, Lemma 9.1]) shows that in fact K does not involve differentiation in any of its variables.

LEMMA 1.5 *The curvature $F(X,Y)Z$ at a point $m \in M$ depends only on the values of X, Y and Z at the point m, and not on their values at nearby points.*

Thus K is induced by a vector-bundle map $TM \otimes TM \to \mathrm{End}(V)$, in other words a tensor field. Observe that $K(X,Y)$ is antisymmetric in X and Y and that we may therefore regard K as a 2-form on M with values in $\mathrm{End}(V)$. Specifically, in terms of local coordinates x^i we define the *curvature 2-form* to be the $\mathrm{End}(V)$-valued 2-form

$$K = \sum_{i<j} K(\partial_i, \partial_j) dx^i \wedge dx^j.$$

(See 1.19 below for a more general discussion of differential forms and our conventions regarding the identification of differential forms with antisymmetric tensors.)

Connections on the tangent bundle itself are of particular interest. Such a connection can be expressed in terms of local coordinates; let x^i be such local coordinates, with corresponding basic vector fields ∂_i. Then each of the n endomorphisms Γ_i of Remark 1.2 can be expressed as $n \times n$ matrices. Specifically, write

$$\nabla_i(\partial_j) = \sum_k \Gamma_{ij}^k \partial_k.$$

The *connection coefficients* or *Christoffel symbols* Γ_{ij}^k determine the connection completely, since if $X = \sum X^i \partial_i$ and $Y = \sum Y^j \partial_j$ are two vector fields, the axioms (i) and (ii) in the definition of a connection imply that

$$\nabla_X Y = \sum_{i,j} X^i Y^j_{;i} \partial_j$$

where by definition

$$Y^j_{;i} = \partial_i Y^j + \sum_a \Gamma_{ia}^j Y^a.$$

Thus, if M is n-dimensional, a connection is given locally by the choice of the n^3 functions Γ_{ij}^k.

REMARK 1.6 In the above discussion the local coordinates enter only through the framing of the tangent bundle provided by $\{\partial_i\}$. While it is most usual to work in terms of coordinate framings, we will also have occasion to use other framings of the tangent bundle. The discussion above can also be carried out with respect to a general framing.

11

DEFINITION 1.7 A connection on TM is said to be *symmetric* (or *torsion-free*) if for any two vector fields X and Y, with Poisson bracket $[X,Y]$,

$$\nabla_X Y - \nabla_Y X = [X,Y].$$

In terms of the Christoffel symbols, this condition is equivalent to $\Gamma^k_{ij} = \Gamma^k_{ji}$ in any coordinate framing.

Riemannian geometry

Now suppose that M is a Riemannian manifold, that is, M is equipped with a smoothly varying choice of inner product on the fibers of TM. We denote these inner products $(\ ,\ )$.

DEFINITION 1.8 One says that ∇ is *compatible with the metric* if for any three vector fields X, Y_1, and Y_2,

$$(\nabla_X Y_1, Y_2) + (Y_1, \nabla_X Y_2) = X \cdot (Y_1, Y_2).$$

A more geometrical statement of the same condition is that parallel translation preserves inner products (think of X as $\dot\gamma$ to see the relation between these two statements).

THEOREM 1.9 (LEVI-CIVITA) *A Riemannian manifold possesses a unique connection that is symmetric and compatible with the metric.*

PROOF We work in local coordinates. As usual, the metric is represented by a symmetric matrix of smooth functions $g_{jk} = (\partial_j, \partial_k)$. Now the condition of compatibility with the metric implies

$$\partial_i g_{jk} = \sum_a (\Gamma^a_{ij} g_{ak} + \Gamma^a_{ik} g_{aj})$$

and by permuting the suffixes we get

$$\partial_j g_{ki} = \sum_a (\Gamma^a_{jk} g_{ai} + \Gamma^a_{ji} g_{ak}), \qquad \partial_k g_{ij} = \sum_a (\Gamma^a_{ki} g_{aj} + \Gamma^a_{kj} g_{ai}).$$

Combine these and use the assumed symmetry of Γ to get

$$\sum_a \Gamma^a_{ij} g_{ak} = \tfrac{1}{2}(\partial_i g_{jk} + \partial_j g_{ik} - \partial_k g_{ij}).$$

This determines the Γ's uniquely, since g is an invertible matrix. □

From now on, we shall always think of a Riemannian manifold as equipped with this connection, called the *Levi-Civita connection*.

REMARK 1.10 If X is a fixed vector field, then ∇_X defines a linear map from the space of vector fields to itself. Vector fields are of course tensor fields of type $\binom{1}{0}$, and in fact ∇_X extends uniquely to a linear map (preserving type) of the space of all tensor fields to itself, satisfying the relations

(i) $\nabla_X (A \otimes B) = \nabla_X A \otimes B + A \otimes \nabla_X B$;

(ii) ∇_X is equal to the Lie derivative on functions (tensor fields of type $\binom{0}{0}$);

(iii) ∇_X commutes with contractions of covariant and contravariant indices.

We will construct such an extension ∇_X later, using principal bundles. Notice that (i) – (iii) allow us to work out the covariant derivative in local co-ordinates. For instance, if $A = \sum_{i,j} A^{ij} \partial_i \otimes \partial_j$ is a tensor of type $\binom{2}{0}$, then (i) implies that

$$\nabla_k A = \sum_{i,j} A^{ij}{}_{,k} \partial_i \otimes \partial_j,$$

where

$$A^{ij}{}_{,k} = \frac{\partial A^{ij}}{\partial x^k} + \sum_a \Gamma^i_{ka} A^{aj} + \sum_a \Gamma^j_{ka} A^{ia}.$$

If $B = \sum_{i,j} B^i_j \partial_i \otimes dx^j$ is a tensor of type $\binom{1}{1}$, the axioms imply

$$\nabla_k B = \sum_{i,j} B^i_{j,k} \partial_i \otimes dx^j$$

where

$$B^i_{j,k} = \frac{\partial B^i_j}{\partial x^k} + \sum_a \Gamma^i_{ka} B^a_j - \sum_a \Gamma^a_{jk} B^i_a.$$

These formulas can obviously be generalized.

Since the metric uniquely determines the Levi-Civita connection, it also uniquely determines its curvature.

DEFINITION 1.11 The curvature of the Levi-Civita connection on a Riemannian manifold is called the *Riemann curvature operator* (or *tensor*) and is denoted by R. In components relative to a frame $\{e_i\}$ for the tangent bundle we may write

$$R(e_j, e_k) e_l = \sum_i R^i_{ljk} e_i.$$

When e_i is a coordinate frame corresponding to local coordinates x^i, direct computations yield

$$R^i_{ljk} = \frac{\partial \Gamma^i_{kl}}{\partial x^j} - \frac{\partial \Gamma^i_{jl}}{\partial x^k} + \Gamma^m_{kl}\Gamma^i_{jm} - \Gamma^m_{jl}\Gamma^i_{km}.$$

It is sometimes convenient also to work with the 4-covariant version of the Riemann curvature tensor defined by

$$R_{iljk} = (R(e_j, e_k)e_l, e_i);$$

classically one would say that we have used the metric to 'lower the index' i of the Riemann tensor, to obtain the covariant version.

PROPOSITION 1.12 *The Riemann curvature operator (or tensor) R has the following symmetries:*

(i) $R(X,Y)Z + R(Y,X)Z = 0$, *or in components,* $R^i_{ljk} + R^i_{lkj} = 0$. *(This is just the statement, true for any connection, that the curvature is a 2-form.)*

(ii) $R(X,Y)Z + R(Y,Z)X + R(Z,X)Y = 0$ *(in components* $R^i_{ljk} + R^i_{jkl} + R^i_{klj} = 0$*). This is the first Bianchi identity, and it depends only on the fact that the connection is torsion-free.*

(iii) $(R(X,Y)Z, W) + (R(X,Y)W, Z) = 0$. *(in components* $R_{iljk} + R_{lijk} = 0$.*)*

(iv) *Finally,* $(R(X,Y)Z, W) = (R(Z,W)X, Y)$, *i.e.* $R_{iljk} = R_{jkil}$.

For proof see [54, Lemma 9.3].

(1.13) The Riemann curvature tensor is a pretty complicated geometric object, and it is reasonable to ask for some kind of condensation of the information contained in it. We could try contracting various indices, but the symmetries of the curvature mean that there is only one non-trivial contraction. This is the *Ricci curvature tensor*, which is the bilinear form on TM defined by

$$\text{Ric}(Y,Z) \mapsto \text{tr}\bigl(X \mapsto R(X,Y)Z\bigr)$$

or, in components, $\text{Ric}_{ab} = \sum_i R^i_{aib}$. It is clear from (iv) above that this tensor is symmetric. (Using the metric, any symmetric bilinear form on TM may be expressed as $(Y,Z) \mapsto (Y, LZ)$ for some self-adjoint linear operator Z; the operator obtained

Notation	Interpretation
$R(X,Y)Z$	vector field obtained from X, Y, Z
$R(X,Y)$	(skew-symmetric) endomorphism of TM
R	$\text{End}(TM)$-valued 2-form
$e_i, \hat{e}_j$	framing for TM and dual framing for T^*M
R_{iljk}	$(e_i, R(e_j, e_k)e_l)$ classical tensor form
R_{il}	2-forms; matrix entries of R; $R_{il} = \sum_{j<k} R_{iljk} \hat{e}_j \wedge \hat{e}_k$
Ric_{ab}	Ricci curvature tensor; equals $\sum_i R^i_{aib}$
κ	scalar curvature; equals $\sum_{i,j} R_{ijij}$ if framing orthonormal
R^S	Riemann endomorphism of a Clifford bundle (see chapter 3)

TABLE 1.1. Notation for various interpretations of the Riemann curvature

in this way from Ric is called the *Ricci curvature operator*.) Contracting further, we define the *scalar curvature*, denoted κ, by

$$\kappa = g^{ab} \text{Ric}_{ab},$$

the trace of the Ricci curvature operator. All these interpretations and variants of the Riemann curvature are summarized in table 1.1.

DEFINITION 1.14 A curve γ in a Riemannian manifold M is a geodesic if $\nabla_{\dot{\gamma}} \dot{\gamma} = 0$, or in other words if $\dot{\gamma}$ is parallel along γ.

The geodesic equation is a second-order differential equation, so it has a unique solution (at least for small values of the time parameter) subject to initial values $\gamma(0), \dot{\gamma}(0)$; in other words, there is a unique geodesic segment through a given point of M in a given direction.

The geodesic equation is *isochronous*; if $t \mapsto \gamma(t)$ is a solution, then so is $t \mapsto \gamma(ct)$ for any constant c. From this fact it follows easily that if m is a point of M, the exponential map

$$\exp : U \to M \quad (U \subseteq T_m M),$$

defined where this makes sense by sending a vector $v \in T_m M$ to the value $\gamma(1)$ of the unique geodesic with $\gamma(0) = m, \dot{\gamma}(0) = v$, is in fact defined on an open subset U of $T_m M$ which is star-shaped about the origin. By the inverse function theorem, exp is a diffeomorphism of a neighbourhood of zero in $T_m M$ to a neighbourhood of m in M. Choosing an orthonormal basis for $T_m M$ gives a special co-ordinate system, called a *geodesic co-ordinate system*, for a neighbourhood of m.

PROPOSITION 1.15 *At the origin of a geodesic co-ordinate system, the Christoffel symbols all vanish.*

PROOF We must show that $\nabla_i \partial_j = 0$ at the origin. Since $\nabla_i \partial_j = \nabla_j \partial_i$ by symmetry of the connection, it is enough to prove that $\nabla_X X = 0$ at the origin for all vector fields $X = \sum X^j \partial_j$, X^j being a constant function. But in a geodesic co-ordinate system, the radial lines through the origin are unit speed geodesics; so X is of constant length and tangent to the geodesic through the origin in direction X. By the geodesic equation, $\nabla_X X = 0$. □

REMARK 1.16 In fact it can be shown that, at the origin in a geodesic coordinate system, the metric tensor g_{ij} has a Taylor expansion whose Taylor coefficients are manufactured out of the curvature tensor at the origin and its covariant derivatives. See exercise 1.32.

A Riemannian manifold has a natural metric space structure. If $\gamma: [0,1] \to M$ is a curve on a Riemannian manifold M, we define its *length* by

$$\text{len}(\gamma) = \int_0^1 |\gamma'(t)| dt$$

where the absolute value $|\gamma'(t)|$ of the tangent vector $\gamma'(t)$ is determined by using the Riemannian metric. If p, q are points of M we define their *distance* by

$$d(p,q) = \inf\{\text{len}(\gamma) : \gamma(0) = p, \gamma(1) = q\}.$$

We quote without proof the following standard facts:

PROPOSITION 1.17 *d is a metric defining the topology of M. Moreover, within the domain of a geodesic co-ordinate system, the balls in the metric d around the origin are just the ordinary balls of the same radius in $\mathbb{R}^n$.*

THEOREM 1.18 (HOPF-RINOW) *M is a complete metric space if and only if every geodesic in M can be extended to arbitrary length.*

For proofs, one can consult Milnor [54, Chapter 10].

Differential forms

(1.19) Differential forms are defined as smooth sections of the exterior bundle of the cotangent bundle, $\Lambda^* T^* M$. To do computations, it is convenient to identify a differential m-form α with an antisymmetric tensor A of type $\binom{0}{m}$ by writing it

$$\alpha = \sum_{i_1 < \cdots < i_m} A_{i_1 \cdots i_m} dx^{i_1} \wedge \ldots \wedge dx^{i_m} = \frac{1}{m!} \sum_{i_1, \ldots, i_m} A_{i_1 \cdots i_m} dx^{i_1} \wedge \ldots \wedge dx^{i_m}.$$

Under this convention, the differential form $dx \wedge dy$ on $\mathbb{R}^2$ is identified with the antisymmetric tensor

$$\begin{pmatrix} 0 & -1 \\ 1 & 0 \end{pmatrix}.$$

An important example is the *curvature 2-form*, which is the $\mathrm{End}(V)$-valued 2-form associated to the curvature operator of a connection on V; we may write it

$$K = \sum_{i<j} F(\partial_i, \partial_j) dx^i \wedge dx^j.$$

In particular we have the *Riemann curvature 2-form* which may be regarded as a matrix of 2-forms R_{il} defined by

$$R_{il} = \sum_{j<k} R_{iljk} dx^j \wedge dx^k.$$

If α is an m-form with associated antisymmetric tensor A then define for vectors $X_1, \ldots, X_m$

$$\alpha(X_1, \ldots, X_m) = (X_1 \otimes \cdots \otimes X_m)(A);$$

we may write this in index notation, for instance for $m = 2$:

$$\alpha(X, Y) = \sum_{i,j} X^i Y^j A_{ij}$$

where X^i and Y^j are the components of X and Y. These conventions introduce certain constants into expressions involving differential forms. Other publications — and in particular the first edition of this book — use different conventions and have different constants. You have been warned!

The exterior product of differential forms may be represented in tensorial form by means of the generalized Kronecker delta symbol:

$$\delta^{j_1\cdots j_m}_{i_1\cdots i_m} = \begin{vmatrix} \delta^{j_1}_{i_1} & \cdots & \delta^{j_1}_{i_m} \\ \vdots & \ddots & \vdots \\ \delta^{j_m}_{i_1} & \cdots & \delta^{j_m}_{i_m} \end{vmatrix}.$$

This symbol is equal to +1 if the 'j' indices are distinct and an even permutation of the 'i' indices, to -1 if the 'j' indices are distinct and an odd permutation of the 'i' indices, and to 0 if the 'j' indices and the 'i' indices do not form the same set of n elements.

Now if $\alpha \in \Omega^p$ and $\beta \in \Omega^q$ correspond to antisymmetric tensors A and B, then $\alpha \wedge \beta$ corresponds to the antisymmetric tensor C,

$$C_{k_1\cdots k_{p+q}} = \frac{1}{p!q!} \sum_{i_1,\ldots,i_p,j_1,\ldots,j_q} \delta^{i_1\cdots i_p j_1\cdots j_q}_{k_1\cdots k_{p+q}} A_{i_1\cdots i_p} B_{j_1\cdots j_q}.$$

Similarly, the exterior derivative $d\alpha$ corresponds to the antisymmetric tensor D, where

$$D_{k_1\cdots k_{p+1}} = \frac{1}{p!} \sum_{i_1,\ldots,i_p} \delta^{j i_1\cdots i_p}_{k_1\cdots k_{p+1}} \frac{\partial A_{i_1\cdots i_p}}{\partial x^j}.$$

In this formula we may, if we wish, replace $\partial A_{i_1\cdots i_p}/\partial x^j$ by the covariant derivative $A_{i_1\cdots i_p;j}$; for the difference will be a sum of terms like $\sum_a A_{i_1\cdots i_{q-1} a i_{q+1}\cdots i_p} \Gamma^a_{j i_q}$, which is symmetric in j and i_q and will therefore vanish on antisymmetrization.

DEFINITION 1.20 Let M be an oriented Riemannian n-manifold. Let $x^1,\ldots,x^n$ be oriented local co-ordinates. We define the symbol $g = \det(g_{ij})$. Now we define the *volume form* vol $\in \Omega^n(M)$ by

$$\text{vol} = \sqrt{g}dx^1 \wedge \ldots \wedge dx^n.$$

It is easy to check that this expression is in fact independent of local co-ordinates so that it defines a global n-form on M.

The metric on TM induces a metric on all the associated tensor bundles, and therefore also on $\wedge^k T^*M$, considered as a bundle of antisymmetric tensors. To fix

a definite normalization, if $\alpha, \beta \in \Omega^k$ correspond to antisymmetric tensors A and B, let us set
$$(\alpha, \beta) = \frac{1}{k!} \sum_{i_1,\ldots,i_k \supset_1,\ldots j_k} g^{i_1 j_1} \cdots g^{i_k j_k} A_{i_1 \cdots i_k} B_{j_1 \cdots j_k},$$
where $g^{ij} = (dx^i, dx^j)$ is the inverse matrix to g_{ij}. (This definition is arranged so that $(\mathrm{vol}, \mathrm{vol}) = 1$.)

DEFINITION 1.21 Let α be a k-form. Define $*\alpha$ to be the unique $(n-k)$-form such that for all k-forms β
$$(\alpha, \beta) \, \mathrm{vol} = \beta \wedge *\alpha.$$

The operation $*$ is linear and has the property that $**\alpha = (-1)^{kn+k}\alpha$. Thus $*$ is almost an involution. One can check this simply after choosing orthonormal co-ordinates at a point.

DEFINITION 1.22 If α is a k-form, define
$$d^*\alpha = (-1)^{nk+n+1} * d * \alpha.$$

Thus $d^*\alpha$ is a $(k-1)$-form. Clearly, $(d^*)^2 = 0$. The importance of d^* lies in the fact that it is the *formal adjoint* of d. Specifically, let α and β be forms of the same degree. Define their global inner product by
$$\langle \alpha, \beta \rangle = \int_M (\alpha, \beta) \, \mathrm{vol} = \int_M \beta \wedge *\alpha = \int_M \alpha \wedge *\beta.$$
This makes sense if at least one of α and β is compactly supported. Now

PROPOSITION 1.23 If α, β are smooth forms of degrees k and $k-1$ on the oriented Riemannian manifold M, and one of them is compactly supported, then
$$\langle \alpha, d\beta \rangle = \langle d^*\alpha, \beta \rangle.$$

PROOF By Stokes' theorem
$$\begin{aligned}
0 &= \int_M d(\beta \wedge *\alpha) = \int_M d\beta \wedge *\alpha + (-1)^{k-1} \int_M \beta \wedge d(*\alpha) \\
&= \langle \alpha, d\beta \rangle + (-1)^{k-1+(n-k+1)n+(n-k+1)} \int_M \beta \wedge **d(*\alpha) \\
&= \langle \alpha, d\beta \rangle - \langle d^*\alpha, \beta \rangle. \quad \square
\end{aligned}$$

DEFINITION 1.24 The Laplacian Δ is defined to be $dd^* + d^*d = (d + d^*)^2$.

EXAMPLE 1.25 Let $\alpha = \sum A_i dx^i$ be a 1-form. Then $d^*\alpha$ is a 0-form, i.e. a function, often called the *divergence* of α. (The *divergence theorem* $\int d^*\alpha \cdot \text{vol} = 0$ is just a reformation of Stokes' theorem.) Let us calculate $d^*\alpha$, which is equal to $-*d*\alpha$.

First, it is easy to check that

$$*\alpha = \sum_{i,j} (-1)^{j+1} \sqrt{g} A_i g^{ij} dx^1 \wedge \ldots dx^{j-1} \wedge dx^{j+1} \wedge \ldots dx^n.$$

Therefore

$$d(*\alpha) = \sum_{i,j} \frac{\partial}{\partial x^j} (A_i g^{ij} \sqrt{g}) dx^1 \wedge \ldots \wedge dx^n$$

and

$$d^*\alpha = \frac{-1}{\sqrt{g}} \sum_{i,j} \frac{\partial}{\partial x^j} (A_i g^{ij} \sqrt{g}). \tag{1.26}$$

This expression can be simplified as follows. Write

$$d^*\alpha = \sum_{i,j} \left(-g^{ij} \frac{\partial A_i}{\partial x^j} - A_i \frac{\partial g^{ij}}{\partial x^j} - A_i g^{ij} \frac{\partial}{\partial x^j} \log \sqrt{g} \right).$$

From the compatibility of the metric and connection

$$\sum_j \frac{\partial g^{ij}}{\partial x^j} = -\sum_{a,J} \left(\Gamma^j_{ja} g^{ai} + \Gamma^i_{ja} g^{aj} \right).$$

To evaluate $(\partial/\partial x^j) \log \sqrt{g}$, differentiate the determinant g, remembering that g^{ab} is the cofactor of g_{ab}; this gives

$$\frac{\partial g}{\partial x^j} = \sum_{a,b} g^{ab} g \frac{\partial g_{ab}}{\partial x^j}.$$

and it follows that

$$\frac{\partial}{\partial x^j} \log \sqrt{g} = \sum_a \Gamma^a_{ja}.$$

Putting the pieces together, we get

$$d^*\alpha = -\sum_{i,j} g^{ij} \frac{\partial A_i}{\partial x^j} + \sum_{i,j,k} A_i g^{jk} \Gamma^i_{jk}$$

$$= -\sum_{i,j} g^{ij} A_{i,j}. \tag{1.27}$$

Formulae (1.26) and (1.27) may be applied to $\alpha = df$ to give an explicit formula for the Laplacian of a function f

$$\Delta f = \frac{-1}{\sqrt{g}} \sum_{i,j} \frac{\partial}{\partial x^i} \{ \sqrt{g} g^{ij} \frac{\partial f}{\partial x^j} \} = \sum_{i,j} -g^{ij} f_{,ij}.$$

Notice that if $g^{ij} = \delta^{ij}$, then Δ is just the usual Laplacian of Euclidean space.

Exercises

QUESTION 1.28 Prove *Cartan's formula* for the exterior derivative of $\alpha \in \Omega^p(M)$, namely

$$d\alpha(X_0, \ldots, X_p) = \sum_i (-1)^i \alpha(X_0, \ldots, \widehat{X_i}, \ldots, X_p)$$
$$+ \sum_{i<j} (-1)^{i+j} \alpha([X_i, X_j], X_0, \ldots, \widehat{X_i}, \ldots, \widehat{X_j}, \ldots, X_p)$$

where $X_0, \ldots, X_p$ are vector fields, and the 'hat' denotes omission of the specified term.

QUESTION 1.29 Prove the *second Bianchi identity* (for the Riemann curvature), which states that

$$R^i_{ljk,m} + R^i_{lmj,k} + R^i_{lkm,j} = 0.$$

QUESTION 1.30 Let x^i be a geodesic coordinate system on a manifold M, and let r denote the Riemannian distance from the point x to the origin. Prove that

$$\sum_{i,j} g_{ij}(x) x^i x^j = \sum_{i,j} g^{ij}(x) x^i x^j = r^2.$$

QUESTION 1.31 Let α be a k-form on an n-dimensional oriented Riemannian manifold M, represented (in oriented local coordinates) by an antisymmetric tensor A. Show that

$$*\alpha = \sum_{i_1 \ldots i_k j_1 \ldots j_{n-k}} \frac{1}{(n-k)! k!} \delta^{12 \ldots n}_{i_1 \ldots i_k j_1 \ldots j_{n-k}} \sqrt{g} A^{i_1 \ldots i_k} dx^{j_1} \wedge \ldots \wedge dx^{j_{n-k}}.$$

Notice that in this formula we have used the metric to 'raise the indices' of A.)

QUESTION 1.32 Show that, at the origin of a geodesic co-ordinate system, the metric has the following Taylor expansion:

$$g_{ij}(x) = \delta_{ij} + \tfrac{1}{3} \sum_{p,q} x^p x^q R_{ipqj}(0) + O(|x|^3).$$

CHAPTER 2

Connections, curvature, and characteristic classes

Principal bundles and their connections

Let G be a Lie group. Recall that a *principal bundle* E with *structural group* G over a smooth manifold M is a locally trivial fiber bundle whose fiber is G itself considered as a right G-space. Thus, G acts smoothly on E by right multiplication on each fiber and $M = E/G$.

EXAMPLE 2.1 Let V be a k-dimensional vector bundle. The *principal bundle of frames* for V is the space E whose fiber over a point $m \in M$ is the collection of all frames in the fiber V_m of V over m. Clearly, E is a principal bundle with group $GL(k)$. If V has a metric, one can consider similarly the principal bundle of *orthonormal* frames for V, which has structural group $O(k)$ or $U(k)$ according to whether V is real or complex.

Conversely, let E be a principal bundle with group G and let $\rho\colon G \to GL(F)$ be a representation of G on a vector space F. Then G operates on the space $E \times F$ by

$$(e, f)g = (eg, \rho(g^{-1})f)$$

and the quotient space $E \times_\rho F$ of $E \times F$ by this action of G is a vector bundle over M with fibers isomorphic to F. It is called the vector bundle *associated to E by the representation ρ*.

It is worth making a few comments about functions and differential forms on principal bundles. Let E be a principal bundle with group G. Differentiating the G-action we find that to each element u of $\mathfrak{g}$, the Lie algebra of G, there is associated a G-invariant vector field X_u on E, called the *Killing field* corresponding to u. The Killing fields span a subbundle VE of TE, which is equal to the kernel of the map $\pi\colon TE \to TM$ and is called the subbundle of *vertical tangent vectors* to E; each fiber of VE is thus canonically identified with $\mathfrak{g}$. A differential form $\alpha \in \Omega^p(E)$ is called *horizontal* if $\alpha(X_1, \ldots, X_p) = 0$ whenever at least one of the vectors $X_1, \ldots, X_p$

is vertical. If $\beta \in \Omega^p(M)$ is a form it is clear that $\pi^*\beta \in \Omega^p(E)$ is horizontal and invariant. In fact, π^* gives a bijection between the differential forms on M and the horizontal, invariant differential forms on E.

We can generalize this discussion of *invariant* forms to *equivariant* forms. Let (F, ρ) be a representation-space of G as above. The space of functions $f: E \to F$ has two commuting left G-actions, one coming from the right G-action on E and one from the left G-action on F. A function f is called ρ-*equivariant* if it is invariant for the product of these two actions, in other words if

$$\rho(g^{-1})f(e) = f(eg) \quad \forall e \in E, \; g \in G.$$

Similarly we can define ρ-*equivariant differential forms* on E, with values in F.

LEMMA 2.2 *In the above situation there is a 1 : 1 correspondence between ρ-equivariant functions on E and sections of the associated vector bundle $V = E \times_\rho F$. Similarly there is a 1 : 1 correspondence between ρ-equivariant, horizontal differential forms on E and V-valued differential forms on M.*

The proof is simply a matter of chasing definitions, and is left to the reader.

What should be the correct notion of connection on a principal bundle? To answer this question, consider the special case where E is the frame bundle of a vector bundle V. A connection on V is defined by the parallel translation that it induces, and this parallel translation can be thought of as a G-equivariant way of lifting paths from M to E. Differentiating, one gets a way of lifting tangent vectors from M to E.

To put this precisely, let $\pi: E \to M$ be the canonical projection and let $VE = \ker(T_\pi) \leq TE$ be the sub-bundle of vertical tangent vectors. Then there is an exact sequence of vector bundles over E,

$$0 \longrightarrow VE \longrightarrow TE \xrightarrow{T_\pi} \pi^*TM \longrightarrow 0. \tag{2.3}$$

DEFINITION 2.4 We define a *connection on E* to be a G-equivariant choice of splitting for the exact sequence 2.3.

In other words, a connection is a G-equivariant choice of complementary subbundle $HE \leq TE$, the bundle of *horizontal tangent vectors* of the connection, such that $TE = VE \oplus HE$. Our discussion above shows that a connection on a vector bundle V determines a connection on its frame bundle. Similarly, a connection on a Hermitian or Euclidean vector bundle that is compatible with the metric (in the sense that parallel translation preserves inner products) determines a connection on its orthogonal frame bundle.

There are a number of ways to reformulate the definition of a connection. One can also think of the splitting of the exact sequence as given by the induced projection $TE \to VE$. The fibers of VE are naturally identified with the Lie algebra $\mathfrak{g}$ of G, so that this projection can be thought of as a $\mathfrak{g}$-valued 1-form ω on E. A $\mathfrak{g}$-valued 1-form ω on E is a connection 1-form if

(i) it is equivariant: $\omega(\xi.g) = \mathrm{Ad}(g^{-1})\omega(\xi)$;

(ii) it represents a projection: for $u \in \mathfrak{g}$, we must have $\omega(X_u) = u$, where X_u denotes the Killing vector field on E induced by u.

Finally, associated to the direct sum decomposition $TE = VE \oplus HE$ there is a *projection* P_ω of the space of differential forms on E onto the subspace of horizontal forms. The projection P_ω, the connection 1-form ω, and the horizontal subbundle HE all determine one another, and we will use whichever is convenient.

(2.5) Let E be a principal bundle over M with structural group G, and suppose that E is equipped with a connection. It is easy then to see that given a path $\gamma : [0,1] \to M$ and a point $e \in E_{\gamma(0)}$, there is a unique path $\tilde{\gamma} : [0,1] \to E$ starting at e such that $\pi \circ \tilde{\gamma} = \gamma$ and that $\tilde{\gamma}'(t)$ is horizontal; call $\tilde{\gamma}$ the *horizontal lift* of γ.

Now let $\rho \colon G \to GL(F)$ be a representation of G on a vector space F, and let $V = E \times_G F$ be the associated vector bundle. Given a vector $w_0 \in W_{\gamma_0}$, represented as $w_0 = (e, f)$, define the *parallel translate* of w_0 along γ to be the vector $w_1 \in W_{\gamma(1)}$ given by $w_1 = (\tilde{\gamma}(1), f)$, where $\tilde{\gamma}$ is (as above) the horizontal lifting of γ starting at e. Using the equivariance, it is easy to check that the vector w_1 does not depend on the choice of the representation of w_0 as (e, f).

We can use this parallel translation to define a connection in the vector bundle W.

25

To do this, let X be a tangent vector at a point $m \in M$, and let w be a section of W. Choose a curve γ so that $\gamma(0) = m$, $\gamma'(0) = X$ and let $\theta_t : W_{\gamma(t)} \to W_{\gamma(0)}$ be the isomorphism induced by parallel translation. Define

$$\nabla_X w = \frac{d}{dt}\{\theta_t w_{\gamma(t)}\}\Big|_{t=0}. \qquad (2.6)$$

PROPOSITION 2.7 *With notation as above:*

(i) *Formula (2.6) is independent of the choice of the curve γ and defines a connection on W.*

(ii) *If we identify sections of W with ρ-equivariant functions $E \to F$, as in lemma 2.2, then ∇_X corresponds to the directional derivative along $\tilde{X}$, the horizontal lift of X;*

(iii) *Equivalently, the operator $\nabla : \Omega^0(V) \to \Omega^1(V)$ corresponds to the map $f \mapsto P_\omega df$ from ρ-equivariant functions on E to horizontal, ρ-equivariant E-valued 1-forms.*

PROOF It is enough to prove (ii), since the other parts are immediate consequences; and for this, notice that if $f : E \to F$ is the equivariant function corresponding to w, so that $w = (e, f(e))$, then by definition of parallel translation

$$\theta_t w_{\gamma(t)} = \theta_t\big(\tilde{\gamma}(t), x(\tilde{\gamma}(t))\big) = \big(e, x(\tilde{\gamma}(t))\big),$$

where $\tilde{\gamma}$ denotes the horizontal lift of γ starting at e. Therefore

$$\frac{d}{dt}\{\theta_t w_{\gamma(t)}\}\Big|_{t=0} = \big(e, dx(\tilde{\gamma}'(0))\big) = \big(e, \tilde{X}.x(e)\big).$$

as required. □

EXAMPLE 2.8 Let V be a vector bundle with fibers isomorphic to F; we have seen that a connection on V gives rise to one on the principal $GL(F)$-bundle of frames for V. Now let

$$\rho : GL(F) \to GL(F \otimes \ldots \otimes F \otimes F^* \otimes \ldots \otimes F^*)$$

be a tensor product representation; then the bundle associated to ρ is the tensor product $V \otimes \ldots \otimes V \otimes V^* \otimes \ldots \otimes V^*$, and the induced connection is the tensor product connection of (1.10).

The operation $P_\omega d$ (taking the horizontal part of the exterior derivative of a form on E) is called the *exterior covariant derivative* on E. Notice that (iii) above shows that it corresponds to the covariant derivative ∇ on sections of an associated bundle.

PROPOSITION 2.9 Let α be a ρ-equivariant horizontal p-form on E. Then its exterior covariant derivative $P_\omega d\alpha$ is given by the formula

$$P_\omega d\alpha = d\alpha + \rho_* \omega \wedge \alpha.$$

To explain the notation here, $\rho_* : \mathfrak{g} \to \mathfrak{gl}(F) = \operatorname{End}(F)$ is the Lie algebra homomorphism induced by ρ, so that $\rho_*\omega$ is an $\operatorname{End}(F)$-valued one-form on E. The wedge product

$$\Omega^1(E; \operatorname{End}(F)) \otimes \Omega^p(E; F) \to \Omega^{p+1}(E; F)$$

is obtained by tensoring the exterior product on ordinary (scalar-valued) forms with the natural pairing $\operatorname{End}(F) \otimes F \to F$.

PROOF We will employ Cartan's formula for the exterior derivative (1.28) to check that both sides are equal on a $(p+1)$-tuple of vector fields $X_0, \ldots, X_p$, where we may assume by linearity and locality that (for some r) the first r of the X_i are Killing fields and the remaining $p+1-r$ are horizontal and G-invariant. We distinguish three cases.

If $r = 0$, so that $X_0, \ldots, X_p$ are all horizontal, then equality results from the definition of P_ω and the vanishing of ω on horizontal vector fields.

If $r \geqslant 2$ so that X_0 and X_1 are Killing fields, then every term in Cartan's formula has at least one vertical vector field as argument; so $d\alpha(X_0, \ldots, X_p) = 0$ and both sides are zero.

If $r = 1$ so that X_0 is a Killing field and $X_1, \ldots, X_p$ are horizontal and G-invariant, then Cartan's formula gives

$$d\alpha(X_0, \ldots, X_p) = X_0 \cdot \alpha(X_1, \ldots, X_p)$$

since all other terms vanish. But the ρ-equivariance of α gives

$$X_0 \cdot \alpha(X_1, \ldots, X_p) + \rho_*(\omega(X_0))(\alpha(X_1, \ldots, X_p)) = 0$$

and this proves the result in this case also. $\square$

Now let ω be the $\mathfrak{g}$-valued 1-form of a connection on the principal bundle E.

DEFINITION 2.10 We define the *curvature* Ω of ω to be the $\mathfrak{g}$-valued 2-form

$$\Omega(X_1, X_2) = d\omega(X_1, X_2) + [\omega(X_1), \omega(X_2)]$$

where $[.,.]$ refers to the Lie bracket in $\mathfrak{g}$.

REMARK 2.11 Suppose that $\mathfrak{g}$ is a Lie algebra of matrices (that is, a subalgebra of $\mathfrak{gl}(n)$). Matrix-valued differential forms can be made into an (associative) algebra, with a product which is obtained by combining the exterior product on forms with the usual multiplication of matrices. The formula above may then be written $\Omega = d\omega + \omega \wedge \omega$.

The curvature represents the square of the exterior covariant derivative.

PROPOSITION 2.12 For any horizontal, ρ-equivariant p-form α on E, one has

$$P_\omega dP_\omega d\alpha = \rho_* \Omega \wedge \alpha.$$

In particular, Ω itself is a horizontal and Ad-equivariant form on E.

PROOF By 2.9, on horizontal and equivariant forms,

$$P_\omega d\alpha = d\alpha + \rho_* \omega \wedge \alpha.$$

Applying 2.9 again gives

$$P_\omega dP_\omega d\alpha = \rho_*\omega \wedge d\alpha + d(\rho_*\omega \wedge \alpha) + \rho_*\omega \wedge \rho_*\omega \wedge \alpha = (\rho_*(d\omega) + \rho_*\omega \wedge \rho_*\omega) \wedge \alpha.$$

The result follows from the definition of curvature. □

REMARK 2.13 Notice that $\rho_*\Omega$ corresponds to a 2-form on M with values in $\text{End}(V)$ where $V = E \times_\rho F$ is the vector bundle associated to the representation ρ. It is easy to check that this 2-form is just the curvature 2-form K as defined in chapter 1 of the linear connection ∇ on V associated to ω in 2.6. Indeed, one has from the definition and Cartan's formula for the exterior derivative

$$\rho_*\Omega(X, Y)v = (\nabla_X \nabla_Y - \nabla_Y \nabla_X - \nabla_{[X,Y]})v$$

and this is exactly the definition of the curvature 2-form.

EXAMPLE 2.14 Let E be the principal bundle of frames of a m-dimensional vector bundle V. A *framing* of V is a section s of E; such framings exist locally. Using s we may pull back forms on E to forms on M, so that a connection on E is defined by the End(V)-valued 1-form $s^*\omega$ and its curvature by the End(V)-valued 2-form $s^*\Omega$. It follows from 2.9 that, if we use the framing s to trivialize V locally (so that sections of V are represented by $\mathbb{R}^m$-valued functions), then relative to this trivialization we can write the connection as

$$\nabla = d + s^*\omega.$$

Suppose now that we also choose local coordinates x^i; then we may write

$$s^*\omega = \sum_i \Gamma_i dx^i$$

where the Γ_i are End(V)-valued functions. In fact, these functions are simply the Christoffel symbols of Chapter 1.

DEFINITION 2.15 The framing s of V is called *synchronous* near p (relative to the given local coordinates) if s is parallel along radial lines emanating from the origin p.

Clearly such framings can be obtained by choosing any framing over p itself and then extending by parallel transport along radial lines. If W has a metric with which the connection is compatible, then such a framing may be chosen to be orthonormal. By definition, we have

PROPOSITION 2.16 *At the origin of a synchronous framing the Christoffel symbols all vanish.*

In fact we have a Taylor expansion, analogous to 1.32; see exercise 2.33.

Characteristic classes

The theory of characteristic classes comes from the simple question: How can we tell two vector bundles apart? For instance, how do we know that the tangent bundle to the 2-sphere is non-trivial? (An elementary proof uses the residue theorem.) Characteristic classes give a systematic approach.

DEFINITION 2.17 A *characteristic class* c is a natural transformation which to each vector bundle V over a manifold M associates an element $c(V)$ of the cohomology group $H^*(M)$, with the property that if $V_1 \cong V_2$ then $c(V_1) = c(V_2)$.

The word "natural" can be given a precise sense by means of category theory. Of course, in the above definition the bundles may be real or complex, the cohomology may be taken with various coefficients, and so on.

There are many approaches to the theory of characteristic classes. In this chapter, we will develop the so-called *Chern-Weil* method. We will consider characteristic classes with *complex* coefficients of *complex* vector bundles. We will represent $H^*(M)$ as de Rham cohomology, that is closed forms (the kernel of d) modulo exact forms (the image of d).

The idea of the Chern-Weil method is the following. Suppose that our bundle V is equipped with a connection. In some sense, the curvature of this connection measures the local deviation of V from flatness. Now if V is flat, and the base manifold M is simply-connected, then V is trivial. This suggests that there may be a link between curvature and characteristic classes, which measure the global deviation of V from triviality. Such a link is provided by the theory of invariant polynomials.

DEFINITION 2.18 Let $\mathfrak{gl}_m(\mathbb{C})$ denote the Lie algebra of $m \times m$ matrices over $\mathbb{C}$. An *invariant polynomial* on $\mathfrak{gl}_m(\mathbb{C})$ is a polynomial function $P : \mathfrak{gl}_m(\mathbb{C}) \to \mathbb{C}$ such that for all $X, Y \in \mathfrak{gl}_m(\mathbb{C})$, $P(XY) = P(YX)$. An *invariant formal power series* is a formal power series over $\mathfrak{gl}_m(\mathbb{C})$ each of whose homogeneous components is an invariant polynomial.

For example, the determinant and the trace are invariant polynomials.

LEMMA 2.19 *The ring of invariant polynomials on $\mathfrak{gl}_m(\mathbb{C})$ is a polynomial ring generated by the polynomials $c_k(X) = (-2\pi i)^{-k} \operatorname{tr}(\bigwedge^k X)$, where $\bigwedge^k X$ denotes the transformation induced by X on $\bigwedge^k \mathbb{C}^m$.*

PROOF Let P be any invariant polynomial. If we first of all look at the restriction of P to diagonal matrices, we see that P must be a polynomial function of the diagonal entries. Since these diagonal entries can be interchanged by conjugation, P

must in fact be a *symmetric* polynomial function. Now since P is invariant under conjugation, it must be a symmetric polynomial function of the eigenvalues for all matrices with distinct eigenvalues; since by elementary linear algebra such matrices are conjugate to diagonal matrices. But the set of such matrices is dense in $\mathfrak{gl}_m(\mathbb{C})$, so a continuity argument shows that P is just a symmetric polynomial function in the eigenvalues. Now it is easy to see that $\operatorname{tr}(\wedge^k X)$ is the k-th elementary symmetric function in the eigenvalues of X. The main theorem on symmetric polynomials (Lang, [46, Chapter IV]) states that the ring of symmetric polynomials is itself a polynomial ring generated by the elementary symmetric functions, and this now completes the proof. □

Now let V be a complex vector bundle over M, with connection ∇ and curvature K, which is a 2-form on M with values in $\operatorname{End}(V)$. Choosing a local framing for V, we may identify K with a matrix of ordinary 2-forms. If P is an invariant polynomial, we may apply P to this matrix to get an even-dimensional differential form $P(K)$. Because of the invariant nature of P, the form $P(K)$ is independent of the choice of local framing, and is therefore globally defined.

In terms of the principal $GL_m(\mathbb{C})$-bundle E associated to V, this construction may be phrased as follows. Let Ω be the curvature form of the induced connection on E; Ω is a horizontal, equivariant 2-form on E with values in $\mathfrak{gl}_m(\mathbb{C})$, so $P(\Omega)$ is a horizontal invariant form on E. Such a form is the lift to E of a form on M, and this form on M is $P(K)$.

Whichever approach is adopted, notice that since 2-forms are nilpotent elements in the exterior algebra, all formal power series with 2-form-valued variables in fact converge. Thus, the construction makes good sense if P is merely an invariant formal power series.

PROPOSITION 2.20 *For any invariant polynomial (or formal power series) P, the differential form $P(K)$ is closed, and its de Rham cohomology class is independent of the choice of connection ∇ on V.*

PROOF For the purposes of this proof let us describe an invariant formal power series P as *respectable* if the conclusion of the proposition holds for P. Clearly the sum and product of respectable formal power series are respectable. Thus, it is enough to prove that the generators defined in (2.19) are respectable. Equivalently, since

$$\det(1 + qK) = \sum q^k \operatorname{tr}(\textstyle\bigwedge^k K),$$

it is enough to prove that $\det(1+qK)$, considered as a formal power series depending on the parameter q, is respectable.

If P is a respectable formal power series with constant term a, and g is a function holomorphic about a, then $g \circ P$ is also a respectable formal power series. Hence, $\det(1 + qK)$ is respectable if and only if $\log \det(1 + qK)$ is respectable. We will now prove directly that $\log \det(1 + qK)$ is respectable.

For this purpose we will work in the associated principal bundle E of frames for V, with matrix-valued connection 1-form ω and corresponding curvature 2-form Ω. We will use the formula (2.10)

$$\Omega = d\omega + \omega^2 \qquad (**)$$

where the product in the ring of matrix-valued forms is obtained by tensoring exterior product and matrix multiplication. Now suppose that ω depends on a parameter t; then Ω also depends on t, and if we use a dot to denote differentiation with respect to t, then

$$\dot\Omega = d\dot\omega + \omega\dot\omega + \dot\omega\omega.$$

Consider

$$\begin{aligned}\frac{d}{dt}\log\det(1+q\Omega) &= q\operatorname{tr}\{(1+q\Omega)^{-1}\dot\Omega\}\\ &= \sum_{l=0}^{\infty}(-1)^l q^{l+1}\operatorname{tr}\{\Omega^l(d\dot\omega+\omega\dot\omega+\dot\omega\omega)\}.\end{aligned}$$

We need the *second Bianchi identity*

$$d\Omega = \Omega\omega - \omega\Omega$$

which follows from $(**)$ on taking the exterior derivative and then substituting back for $d\omega$.

We have

$$\begin{aligned}\operatorname{tr}\{\Omega^l(\omega\dot\omega+\dot\omega\omega)\} &= \operatorname{tr}\{\Omega^l\omega\dot\omega - \omega\Omega^l\dot\omega\} \quad \text{(by the symmetry of trace)}\\ &= \operatorname{tr}\{(d\Omega^l)\dot\omega\} \quad \text{(by the Bianchi identity)}.\end{aligned}$$

Therefore
$$\operatorname{tr}\{\Omega^l(d\dot\omega + \omega\dot\omega + \dot\omega\omega)\} = d\operatorname{tr}\{\Omega^l\dot\omega\},$$
so
$$\frac{d}{dt}\log\det(1+q\Omega) = d\sum_{l=0}^{\infty}(-1)^l q^{l+1}\operatorname{tr}\{\Omega^l\dot\omega\}$$
is an exact form; in fact it is the exterior derivative of a horizontal and invariant form on E. Therefore, the projection to the base manifold $(d/dt)\log\det(1+qK)$ is also exact. Now the result follows; for since any connection can be deformed locally to flatness, we see that $\log\det(1+qK)$ is locally exact, that is closed; and since any two connections can be linked by a (differentiable) path, the cohomology class of $\log\det(1+qK)$ is independent of the choice of connection. □

It follows from the proposition that any invariant formal power series P defines a characteristic class for complex vector bundles, by the recipe "pick any connection and apply P to the curvature".

DEFINITION 2.21 The generators c_k defined in 2.19 correspond to characteristic classes called *Chern classes*.

From 2.19, any characteristic class defined by an invariant polynomial is therefore a polynomial in the Chern classes.

(2.22) Suppose now that V is a *real* vector bundle, and let $V_{\mathbf{C}}$ denote its complexification, $V_{\mathbf{C}} = V \otimes_{\mathbf{R}} \mathbf{C}$. The odd Chern classes of $V_{\mathbf{C}}$ are then equal to zero (in complex cohomology). To see this, notice that we can give V a metric and compatible connection. The curvature F of such a connection is skew ($\mathfrak{o}(m)$-valued), so
$$\operatorname{tr}(\bigwedge^k F) = (-1)^k \operatorname{tr}(\bigwedge^k F).$$
The non-vanishing Chern classes of $V_{\mathbf{C}}$ are called the *Pontrjagin classes* of V and are denoted
$$p_k(V) = (-1)^k c_{2k}(V_{\mathbf{C}}).$$
If V is an oriented even-dimensional real vector bundle it also has an extra characteristic class called the *Euler class*, corresponding to the *Pfaffian* invariant polynomial on $\mathfrak{o}(m)$. This will be discussed in the exercises.

33

Genera

Holomorphic functions can be used to build important combinations of characteristic classes. In fact, let $f(z)$ be any function holomorphic near $z = 0$. We can use f to construct an invariant formal power series Π_f by putting

$$\Pi_f(X) = \det(f(\frac{-1}{2\pi i}X));$$

the associated characteristic class is called the *Chern f-genus*. Notice that the Chern f-genus has the properties

(i) for a complex line bundle L, $\Pi_f(L) = f(c_1(L))$;

(ii) for any complex vector bundles V_1 and V_2, $\Pi_f(V_1 \oplus V_2) = \Pi_f(V_1)\Pi_f(V_2)$.

(To see (ii) one uses a direct sum connection.) In fact, it can be seen that these two properties determine the characteristic class Π_f uniquely: this follows from the *splitting principle*[1], which says that given any complex vector bundle V over M, there exist a space X and a map $g: X \to M$ such that g^*V splits as a direct sum of line bundles, and such that $g^*: H^*(M) \to H^*(X)$ is injective. The splitting principle allows one to conclude that a characteristic class is determined by its values on direct sums of line bundles.

If the eigenvalues of the matrix $\frac{-1}{2\pi i}X$ are denoted (x_j), then

$$\Pi_f(X) = \prod f(x_j)$$

is a symmetric formal power series in the x_j, which can therefore be expressed in terms of the elementary symmetric functions of the x_j. But these elementary symmetric functions are just the Chern classes. Thus in the literature the genus $\Pi_f(V)$ is often written as $\Pi_f(V) = \prod f(x_j)$, where $x_1, \ldots, x_m$ are 'formal variables' subject to the relations $x_1 + \cdots + x_m = c_1$, $x_1 x_2 + \cdots + x_{m-1} x_{m-2} = c_2$, and so on. In terms of the splitting principle, the formal variables x_j can be considered to represent the first Chern classes of the line bundles into which g^*V splits.

EXAMPLE 2.23 The *total Chern class*

$$c(V) = 1 + c_1(V) + c_2(V) + \cdots$$

---

[1] We will not formally require the splitting principle, which belongs to a different approach to characteristic class theory, and so we do not give the proof.

is the genus associated with $f(z) = 1 + z$. The multiplicative law $c(V_1 \oplus V_2) = c(V_1)c(V_2)$ is the so-called *Whitney sum* formula for the Chern classes.

EXAMPLE 2.24 The genus associated with $f(z) = (1+z)^{-1}$ can be worked out by expanding the product $\prod(1+x_j)^{-1}$ as

$$(1 - x_1 + x_1^2 - \cdots)(1 - x_2 + x_2^2 - \cdots)(\cdots) = 1 - c_1 + (c_1^2 - c_2) + \cdots$$

If $V_1 \oplus V_2$ is trivial, this expresses the Chern classes of V_2 in terms of those of V_1.

(2.25) The *Chern character* ch is the characteristic class associated to the formal power series $X \mapsto \operatorname{tr} \exp(\frac{-1}{2\pi i} X)$. In terms of the formal variables x_j introduced above we have

$$\operatorname{ch}(V) = \sum e^{x_j}.$$

The Chern character is not a genus in the sense described above, because of the appearance of a sum rather than the product; but it does have the analogous property $\operatorname{ch}(V_1 \oplus V_2) = \operatorname{ch}(V_1) + \operatorname{ch}(V_2)$. Moreover, the identity $e^{z_1} e^{z_2} = e^{z_1+z_2}$ implies that $\operatorname{ch}(V_1 \otimes V_2) = \operatorname{ch}(V_1)\operatorname{ch}(V_2)$. Thus, ch is a kind of "ring homomorphism". Direct calculation of the first few terms yields

$$\operatorname{ch}(V) = (\dim V) + c_1 + \tfrac{1}{2}(c_1^2 - 2c_2) + \cdots$$

(2.26) There is an analogous theory of genera for *real* vector bundles. Let g be holomorphic near 0, with $g(0) = 1$. Let f be the branch of

$$z \mapsto \left(g(z^2)\right)^{\frac{1}{2}}$$

which has $f(0) = 1$. Notice that f is an even function of z and therefore the associated genus involves only the even Chern classes. By definition, the *Pontrjagin g-genus* of a real vector bundle V is the Chern f-genus of its complexification. The appearance of the various squares and square roots is explained by the following lemma.

LEMMA 2.27 *Let g be as above. Then for a real vector bundle V, the Pontrjagin g-genus is equal to*

$$\prod_j g(y_j)$$

where the Pontrjagin classes of V are the elementary symmetric functions in the formal variables y_j.

PROOF Regard this as an identity between invariant polynomials over $\mathfrak{o}(n)$. Any matrix in $\mathfrak{o}(n)$ is similar to one in block diagonal form, where the blocks are 2×2 and are of the form
$$X = \begin{pmatrix} 0 & \lambda \\ -\lambda & 0 \end{pmatrix}$$
with eigenvalues $\pm i\lambda$. Since both sides of the desired identity are multiplicative for direct sums, it is enough to prove it for this block X. Now
$$c_1(X) = 0, \quad c_2(X) = \frac{1}{(2\pi i)^2}(i\lambda)(-i\lambda) = -\frac{\lambda^2}{4\pi^2}.$$
Thus $y = p_1(X) = \lambda^2/4\pi^2$. On the other hand, X is similar over $\mathbb{C}$ to
$$\begin{pmatrix} -i\lambda & 0 \\ 0 & i\lambda \end{pmatrix},$$
and so
$$\Pi_f(X) = f(\frac{-\lambda}{2\pi})f(\frac{\lambda}{2\pi}) = (f(\lambda/2\pi))^2 = g(\lambda^2/4\pi^2) = g(y)$$
as required. □

As in the complex case, one can also interpret this lemma in terms of an appropriate splitting principle; one can take a suitable pull-back of V to split as a direct sum of real 2-plane bundles, and the y_j are then the first Pontrjagin classes of the summands

EXAMPLE 2.28 Two important examples are the $\hat{A}$-genus $\hat{A}(V)$, which is the Pontrjagin genus associated to the holomorphic function
$$z \mapsto \frac{\sqrt{z}/2}{\sinh \sqrt{z}/2},$$
and Hirzebruch's $\mathcal{L}$-genus $\mathcal{L}(V)$, which is the Pontrjagin genus associated with the holomorphic function
$$z \mapsto \frac{\sqrt{z}}{\tanh \sqrt{z}}.$$
As we will see, these combinations of characteristic classes arise naturally in the proof of the Index Theorem.

Notes

A comprehensive reference for the geometry of principal bundles and associated connections is Kobayashi and Nomizu [44]. Volume II of this book also contains an account of Chern-Weil theory. Milnor and Stasheff [55] give a thorough discussion of characteristic classes. One should consult this book to learn about the relation between the Chern-Weil method and other approaches to characteristic class theory which show in particular that the Chern and Pontrjagin classes belong to the cohomology with *integer* coefficients, a fact which has important topological implications.

Exercises

QUESTION 2.29 Let ω, Ω be the connection and curvature forms of a connection on a principal bundle. Prove that if X and Y are horizontal vector fields, then $\omega([X,Y]) = -\Omega(X,Y)$.

QUESTION 2.30 Let E be a principal bundle equipped with a connection. One says that E is *flat* if the curvature is zero. Prove E is flat if and only if the horizontal sub-bundle of TE is integrable (that is, tangent to a foliation).

QUESTION 2.31 Let G be a Lie group and H a closed subgroup; consider G as the total space of a principal H-bundle with base the coset space G/H. Suppose that there is a direct sum decomposition $\mathfrak{g} = \mathfrak{h} \oplus \mathfrak{m}$, where $\mathfrak{m}$ is an H-invariant subspace of the Lie algebra $\mathfrak{g}$.

Prove that the $\mathfrak{h}$-component of the canonical $\mathfrak{g}$-valued one-form on G (the Maurer-Cartan 1-form) determines a left G-invariant connection on the bundle. Show that for $X, Y \in \mathfrak{m}$ the curvature of this connection is given by $\Omega(X,Y) = -[X,Y]_\mathfrak{h}$, where the subscript denotes the $\mathfrak{h}$-component in the direct sum decomposition.

QUESTION 2.32 This exercise considers the relation between connections on a vector bundle V and its dual V^*.

(i) Suppose that a connection on V has curvature K, and End(V)-valued 2-form. Prove from the definitions that the End(V^*)-valued curvature 2-form of V^*, for the associated connection, is $-K^*$, where K^* denotes the dual endomorphism to K.

(ii) Give an alternative proof of this by considering V^* as the bundle associated to the frame bundle of V by the contragredient representation $g \mapsto (g^{-1})^t: GL(n) \to GL(n)$.

(iii) Now suppose that V is provided with a Euclidean metric and that the connection is compatible with this metric. Show that the curvature operators of V and V^* agree under the identification of V with V^* provided by the metric.

QUESTION 2.33 Prove that at the origin of a synchronous framing for a vector bundle V there is a Taylor expansion for the Christoffel symbols

$$\Gamma_j = -\tfrac{1}{2}\sum_{j,k} K(\partial_j, \partial_k)x^k + O(|x|^2)$$

where K is the curvature operator of the connection on V.

QUESTION 2.34 Let M denote complex projective n-space $\mathbb{CP}^n$. Let V be the canonical complex line bundle over M.

(i) Assume (or prove if you wish) that $H^*(\mathbb{CP}^n)$ is a truncated polynomial ring on $a = c_1(V)$ as generator. (See [55]).

(ii) Let T denote the tangent bundle to M, considered as a complex vector bundle. Prove that $T \oplus \mathbb{C} = \overline{V} \oplus \ldots \oplus \overline{V}$, the direct sum of $(n+1)$ copies of the dual of V. Deduce that the the total Chern class $c(T) = (1-a)^{n+1}$.

(iii) Now let $T_{\mathbb{R}}$ denote the tangent bundle considered as a real bundle. Show that $T_{\mathbb{R}} \otimes \mathbb{C} = T \oplus \overline{T}$. Deduce that the total Pontrjagin class $p(T_{\mathbb{R}}) = (1+a^2)^{n+1}$.

QUESTION 2.35 A (vector) *superbundle* is a vector bundle E provided with a direct sum decomposition $E = E_+ \oplus E_-$; its *super Chern character* is by definition $\text{ch}(E_0) - \text{ch}(E_1)$. The endomorphisms of E form a superalgebra (see 4.1 for the relevant definitions here). If $T = \begin{pmatrix} a & b \\ c & d \end{pmatrix}$ is an endomorphism of E we define $\text{tr}_s(T) = \text{tr}(a) - \text{tr}(d)$.

(i) Prove that tr_s is a *supertrace*, that is, it vanishes on supercommutators.

(ii) Quillen [60] defined a *superconnection* on E to be an odd parity first order differential operator A on the superalgebra $\Omega^*(E)$ of E-valued differential forms, which satisfies the graded version of Liebniz' rule, namely
$$A(\alpha \wedge \theta) = d\alpha \wedge \theta + (-1)^k \alpha \wedge A\theta$$
for $\alpha \in \Omega^k(M)$ and $\theta \in \Omega^*(E)$. Show that the operator A^2 is local, and hence is given by multiplication by an even differential form $K_A \in \Omega^*(E)$. Show that, for any superconnection A, the differential form $\text{tr}_s(\exp(-K_A/2\pi i))$ is closed and represents the super Chern character of E.

QUESTION 2.36 Let V be a real $2m$-dimensional oriented inner product space and let K be an element of $\mathfrak{o}(V)$ (i.e. a skew-adjoint endomorphism of V). Define an element α of $\wedge^2 V$ by $\alpha = \sum_{i<j}(Ke_i, e_j)e_i \wedge e_j$, where (e_i) is an oriented orthonormal basis of V. Note that the exterior power α^m lies in the 1-dimensional space $\wedge^{2m} V$; define the *Pfaffian* $\text{Pf}(K)$ by
$$\alpha^m = m! \text{Pf}(K) e_1 \wedge \ldots \wedge e_{2m}.$$

(i) Show that $\text{Pf}(K)$ does not depend on the choice of basis made in its definition.
(ii) Show that for any $A \in \mathfrak{gl}(V)$, $\text{Pf}(A^t K A) = \det(A)\text{Pf}(K)$, and deduce that Pf is an invariant polynomial on $\mathfrak{o}(2m)$.
(iii) Prove that $\text{Pf}(K)^2 = \det(K)$.
(iv) If K is the curvature of an oriented real vector bundle W with metric and compatible connection, verify that $\text{Pf}(-K/2\pi)$ defines a characteristic class, called the *Euler class* $e(W)$.
(v) Show that if W is the oriented real 2-plane bundle underlying a complex line bundle L, then $e(W) = c_1(L)$.
(vi) Give an extended interpretation of the formal variables y_j introduced in 2.27, so that $e(V) = \prod \sqrt{y_j}$.

QUESTION 2.37 Let E be a complex vector bundle of fiber dimension k, and let $L = \wedge^k E$ be the 'determinant line bundle' of E (whose fiber at any point is the top exterior power of the corresponding fiber of E). Prove that $c_1(E) = c_1(L)$.

CHAPTER 3
Clifford algebras and Dirac operators

Operators such as the Laplacian Δ introduced in the first chapter will be our chief objects of study in this book. Δ is a second-order operator, and it is an important and from some points of view surprising fact that Δ is the square of the first-order operator $D = d + d^*$. The mechanism lying behind this is the theory of Clifford algebras, which will be developed in this chapter.

Clifford bundles and Dirac operators

DEFINITION 3.1 *Let V be a vector space equipped with a symmetric bilinear form, denoted $(\, , \,)$. A* Clifford algebra *for V is by definition a unital algebra A which is equipped with a map $\varphi \colon V \to A$ such that $\varphi(v)^2 = -(v,v)1$, and which is universal among algebras equipped with such maps; that is, if $\varphi' \colon V \to A'$ is another map from V to an algebra and satisfies $\varphi'(v)^2 = -(v,v)1$, then there is a unique algebra homomorphism $A \to A'$ fitting into a commutative diagram*

For example, if the bilinear form is identically zero, then the exterior algebra $\bigwedge^* V$ is a Clifford algebra.

PROPOSITION 3.2 *For any V, a Clifford algebra exists and is unique up to isomorphism.*

PROOF The uniqueness follows by abstract nonsense from the universal property. To construct a Clifford algebra, choose a basis $e_1, \ldots, e_n$ for V, and take A to be spanned by the 2^n possible products $\varphi(e_1)^{k_1}, \ldots, \varphi(e_n)^{k_n}$, each k being either 0 or 1, with multiplication determined by the rule

$$\varphi(v_1)\varphi(v_2) + \varphi(v_2)\varphi(v_1) = -2(v_1, v_2) \quad \square$$

The unique (up to isomorphism) Clifford algebra for V will be denoted $\text{Cl}(V)$. Notice that if $\dim V = n$, then $\dim \text{Cl}(V) = 2^n$. The natural map $\varphi \colon V \to \text{Cl}(V)$ is injective; as a result, one usually identifies $v \in V$ with its image $\varphi(v) \in \text{Cl}(V)$, considering V as a subspace of its own Clifford algebra.

We will now work out a simple special case of the factorization of the Laplacian. Let V be a real inner product space, with orthonormal basis $e_1, \ldots, e_n$, and let $\text{Cl}(V)$ be its Clifford algebra. Let S be a vector space which is also a left module over $\text{Cl}(V)$, and let $C^\infty(V; S)$ denote the smooth S-valued functions on V. Each basis element e_i corresponds to a differential operator ∂_i on $C^\infty(V; S)$. Define the *Dirac operator* D on $C^\infty(V; S)$ by

$$Ds = \sum_i e_i(\partial_i s)$$

Let us calculate

$$D^2 s = \sum_{i,j} e_j \partial_j (e_i \partial_i s) = \sum_{i,j} e_j e_i \partial_j \partial_i s = -\sum_i \partial_i^2 s \, ;$$

thus D^2 is equal to the Euclidean Laplacian.

REMARK 3.3 We left it unstated above whether S should be thought of as a real or a complex vector space. It will be most convenient for us always to consider *complex* modules; thus by a *Clifford module* for a real inner product space V we will mean a left module over the complex algebra $\text{Cl}(V) \otimes_\mathbb{R} \mathbb{C}$, or equivalently a complex vector space S equipped with an $\mathbb{R}$-linear map $c \colon V \to \text{End}_\mathbb{C}(S)$ such that $c(v)^2 = -(v,v)\mathbf{1}$ for all $v \in V$.

The flat-space construction above can be generalized to a Riemannian manifold. If M is such a manifold then TM is a bundle whose fibers are inner product spaces, so it makes sense to form the bundle of Clifford algebras $\text{Cl}(TM)$. Now let S be a bundle of Clifford modules; i.e. the fiber S_m at $m \in M$ is a left module over $\text{Cl}(T_m M) \otimes \mathbb{C}$. The sections of S are to play the role of the S-valued functions in the preceding example. To differentiate such sections, we need a connection on S. We make some compatibility assumptions, summarized in the next definition:

DEFINITION 3.4 Let S be a bundle of Clifford modules over a Riemannian manifold M. S is a *Clifford bundle* if it is equipped with a Hermitian metric and compatible connection such that

(i) The Clifford action of each vector $v \in T_m M$ on S_m is skew-adjoint, that is, $(v \cdot s_1, s_2) + (s_1, v \cdot s_2) = 0$;

(ii) The connection on S is compatible with the Levi-Civita connection on M, in the sense that $\nabla_X(Ys) = (\nabla_X Y)s + Y \nabla_X s$ for all vector fields X, Y and sections $s \in C^\infty(S)$.

The Clifford bundles that we consider will often be $\mathbb{Z}/2$-graded (or 'superbundles' in the language of question 2.35). This means that S is provided with a direct sum decomposition $S = S_+ \oplus S_-$. In this case we will require that the connection and metric respect the decomposition, and that the Clifford action of a tangent vector v is *odd*, meaning that it maps S_+ to S_- and S_- to S_+.

DEFINITION 3.5 The *Dirac operator* D of a Clifford bundle S is the first order differential operator on $C^\infty(S)$ defined by the following composition:

$$C^\infty(S) \to C^\infty(T^*M \otimes S) \to C^\infty(TM \otimes S) \to C^\infty(S)$$

where the first arrow is given by the connection, the second by the metric (identifying TM and T^*M), and the third by the Clifford action.

Notice that, in the graded case, the Dirac operator is odd; it maps sections of S_+ to sections of S_- and *vice versa*. In terms of a local orthonormal basis e_i of sections of TM, one can write

$$Ds = \sum_i e_i \nabla_i s \ . \tag{3.6}$$

We will develop some general properties of these operators first, and then look at examples. It is helpful to introduce the following terminology

DEFINITION 3.7 Let S be a Clifford bundle and let $K \in \Omega^2(\mathrm{End}(S))$ be a 2-form with values in $\mathrm{End}(S)$. Let e_i be a local orthonormal frame for TM. The endomorphism

$$\mathsf{K} = \sum_{i<j} c(e_i)c(e_j) K(e_i, e_j)$$

of S is called the *Clifford contraction* of K; it does not depend on the choice of frame.

To calculate D^2, choose the local orthonormal frame e_i to be synchronous at some point $m \in M$. Then, at m, $\nabla_i e_j = 0$, and the Lie bracket of e_i and e_j also vanishes at m. Therefore, at m,

$$\begin{aligned} D^2 s &= \sum_{i,j} e_j \nabla_j (e_i \nabla_i s) \\ &= \sum_{i,j} e_j e_i \nabla_j \nabla_i s \\ &= -\sum_i \nabla_i^2 s + \sum_{j<i} e_j e_i (\nabla_j \nabla_i - \nabla_i \nabla_j) s \ . \end{aligned}$$

The two terms in the formula are of different kinds. The first term is the the result of applying a second-order operator analogous to the Laplacian to s; we will write it as $\nabla^* \nabla s$ — the notation will be explained shortly. In the second term, $\nabla_j \nabla_i - \nabla_i \nabla_j = K(e_j, e_i)$ is just the curvature of the connection on S, and is an endomorphism of S. The second term therefore is equal to the Clifford contraction K of the curvature, applied to s. We have thus proved the very important *Weitzenbock formula*

$$D^2 s = \nabla^* \nabla s + \mathsf{K} s \ . \tag{3.8}$$

Why the notation $\nabla^* \nabla$? The point is that ∇ can be thought of as a differential operator from $C^\infty(S)$ to $C^\infty(T^*M \otimes S)$. These bundles are equipped with metrics, so that their spaces of sections have natural inner products. With respect to these inner products, ∇ has a formal adjoint ∇^*; then $\nabla^* \nabla$ is an operator from $C^\infty(S)$ to itself, which I claim is precisely the one appearing in the Weitzenbock formula. To check this we have to work out an expression for ∇^*.

LEMMA 3.9 *The operator $\nabla^* : C^\infty(T^*M \otimes S) \to C^\infty(S)$ is given in terms of local coordinates by the formula*

$$\nabla^*(dx^j \otimes s_j) = -\sum_k g_{jk}(\nabla_j s_k - \Gamma^i_{jk} s_i) \ .$$

Therefore, in a synchronous orthonormal frame (e_i),

$$\nabla^*(\sum_i e_i \otimes s_i) = -\sum_i \nabla_i s_i$$

at the origin.

PROOF Expressions for the formal adjoint of a differential operator are obtained by integration by parts. In our context this takes the following form: wanting to prove that
$$\langle s, \nabla^* \varphi \rangle = \langle \nabla s, \varphi \rangle,$$
where $s \in C^\infty(S)$, $\varphi = dx^j \otimes s_j \in C^\infty(T^*M \otimes S)$, we look at the difference of the local inner products, $(s, \nabla^*\varphi) - (\nabla s, \varphi)$, which is a smooth function on M, and try to prove that it is a divergence. Now

$$(s, \nabla^*\varphi) - (\nabla s, \varphi) = \sum_k \left(-g^{jk}(s, \nabla_j s_k) + g^{jk}\Gamma^i_{jk}(s, s_i) - g^{jk}(\nabla_j s, s_k) \right)$$
$$= \sum_k \left(-g^{jk}\frac{\partial}{\partial x^j}(s, s_k) + g^{jk}\Gamma^i_{jk}(s, s_i) \right) = d^*\omega$$

by (1.27), where ω is the 1-form with components (s, s_i), given in coordinate-free notation by $\omega(X) = (X \otimes s, \varphi)$. □

This result justifies the notation $\nabla^*\nabla$ in the Weitzenbock formula. In particular, notice that $\nabla^*\nabla$ is a *positive operator*: $\langle \nabla^*\nabla s, s \rangle = \|\nabla s\|^2 \geq 0$. Now the Clifford-contracted curvature operator K is a self-adjoint endomorphism of the bundle S, so that it makes sense to ask whether it is positive:

THEOREM 3.10 (BOCHNER) *If the least eigenvalue of K at each point of a compact M is strictly positive, then there are no non-zero solutions of the equation $D^2 s = 0$.*

PROOF By a compactness argument, there is a constant $c > 0$ such that $\langle \mathsf{K}s, s \rangle \geq c\|s\|^2$. But by the Weitzenbock formula (3.8), if $D^2 s = 0$ then
$$\langle \mathsf{K}s, s \rangle = \langle D^2 s, s \rangle - \|\nabla s\|^2 \leq 0 \quad \square$$

A basic fact about the Dirac operator is its formal self-adjointness:

PROPOSITION 3.11 *Let s_1 and s_2 be smooth sections of S, one of which is compactly supported. Then*
$$\langle Ds_1, s_2 \rangle = \langle s_1, Ds_2 \rangle$$

PROOF As in (3.9), we must check that the local expression

$$(Ds_1, s_2) - (s_1, Ds_2)$$

is a divergence. We compute in a synchronous framing e_i:

$$(Ds_1, s_2) - (s_1, Ds_2) = \sum_i (e_i \nabla_i s_1, s_2) - (s_1, e_i \nabla_i s_2)$$
$$= \sum_i (\nabla_i e_i s_1, s_2) - (e_i s_1, \nabla_i s_2) = \sum_i \nabla_i (e_i s_1, s_2) = d^* \omega$$

where ω is the 1-form $\omega(X) = -(Xs_1, s_2)$. □

REMARK 3.12 This result could also have been derived from (3.9) (see the exercises). However we gave this derivation because we will later need to make use of the specific nature of the 1-form ω.

Clifford bundles and curvature

In this section we will take a more careful look at the way the compatibility that we have required between the Levi-Civita connection and the connection on a Clifford bundle restricts the form of the curvature tensor for that bundle, and therefore restricts the possibilities for the term K in the Weitzenbock formula. Suppose that S is a Clifford bundle, and let a local orthonormal framing e_i of the tangent bundle be given. Let K be the curvature operator of S, and let R be the corresponding (Riemannian) curvature operator of TM. Let $c \colon TM \to \mathrm{End}(S)$ denote the Clifford action.

LEMMA 3.13 As endomorphisms of S, we have

$$[K(X,Y), c(Z)] = c(R(X,Y)Z)$$

for any tangent vector fields X, and Y, and Z.

PROOF The identity is a pointwise one, so choose a synchronous framing e_i at a point $p \in M$ and assume that $X = e_i$, $Y = e_j$, and $Z = e_k$ near p. Now from the definition of a compatible connection one easily computes that

$$\nabla_i \nabla_j (e_k \cdot s) = (\nabla_i \nabla_j e_k) \cdot s + e_k \cdot \nabla_i \nabla_j s$$

at the point p — the cross-terms vanish because the framing is synchronous. But (again because the framing is synchronous) the curvature of TM or of S at p is given by $\nabla_i \nabla_j - \nabla_j \nabla_i$ with respect to the appropriate connection; so the result follows immediately. □

This result can be interpreted as saying that Clifford multiplication by Rv is the 'obstruction' to the curvature K being an endomorphism of S in the category of $Cl(TM) \otimes \mathbb{C}$-modules. We can remove this obstruction by a suitable 'correction term'. In the computation of the correction term, we need to recall that if $v = e_a$ is a basis vector, then

$$R(e_i, e_j)v = \sum_l R_{laij} e_l$$

where on the right hand side we have expressed the 4-covariant Riemann curvature tensor with respect to the orthonormal frame given by the e's.

DEFINITION 3.14 For a Clifford bundle S as above, define the *Riemann endomorphism* R^S of S to be the $\text{End}(S)$-valued 2-form

$$R^S(X,Y) = \tfrac{1}{4} \sum_{k,l} c(e_k) c(e_l) (R(X,Y) e_k, e_l).$$

It is easy to see that R^S is independent of the choice of orthonormal basis. It is an $\text{End}(S)$-valued two-form, canonically obtained from the Clifford module structure of S. Now, however, we have

LEMMA 3.15 *As endomorphisms of S, we have*

$$[R^S(X,Y), c(Z)] = c(R(X,Y)Z)$$

for any tangent vectors X, Y, and Z.

PROOF We may assume without loss of generality that $Z = e_a$, $X = e_i$, and $Y = e_j$. Now we have

$$R^S(e_i, e_j) c(e_a) - c(e_a) R^S(e_i, e_j) = \tfrac{1}{4} \sum_{k,l} R_{lkij} c([e_k e_l, e_a]).$$

The commutator $[e_k e_l, e_a]$ vanishes if $k = l$ or if k, l, and a are all distinct. So the only terms that survive are those where $a = k \neq l$ or $a = l \neq k$. By the antisymmetry

47

of the curvature tensor on k, l these two cases make equal contributions, so we get

$$\tfrac{1}{4}\sum_{k,l} R_{lkij} c([e_k e_l, e_a]) = \tfrac{1}{2} \sum_l R_{laij} c(2e_l) = c(R(e_i, e_j)e_a)$$

as required. □

From 3.13 and 3.15 we get

PROPOSITION 3.16 *The curvature 2-form K of a Clifford bundle S can always be written*

$$K = R^S + F^S$$

where R^S is the Riemann endomorphism defined in 3.14 and F^S commutes with the action of the Clifford algebra.

Following [12], we call the Clifford module endomorphism F^S of S the *twisting curvature* of the Clifford bundle.

Corresponding to this more refined analysis of the curvature of S, we can obtain a more refined version of the Weitzenbock formula 3.8. First we will need the following useful calculation:

$$\sum_{i,j,k} R_{lkij} c(e_i e_j e_k) = -2 \sum_a \text{Ric}_{la}\, c(e_a) \qquad (3.17)$$

where Ric denotes the Ricci tensor.

To verify 3.17, note that if i, j and k are distinct indices, then $e_i e_j e_k = e_k e_i e_j = e_j e_k e_i$, and on the other hand, $R_{lkij} + R_{lijk} + R_{ljki} = 0$ by the first Bianchi identity; thus all the terms in the sum on the left hand side with i, j and k distinct will cancel. The terms with $i = j$ vanish because of the antisymmetry of R_{lkij} on i and j, so we are left with the terms with $i = k \neq j$ and the terms with $i \neq k = j$. These are equal, each giving

$$\sum_{i,j} R_{liij} c(e_j) = -\sum_a \text{Ric}_{la}\, c(e_a);$$

the result follows.

PROPOSITION 3.18 *Let S be a Clifford bundle with associated Dirac operator D. Then*

$$D^2 = \nabla^*\nabla + F^S + \tfrac{1}{4}\kappa$$

where $\mathsf{F}^S = \sum_{i<j} c(e_i)c(e_j) F^S(e_i, e_j)$ is the *Clifford contraction of the twisting curvature*, and κ is the scalar curvature of the Riemannian metric.

PROOF Comparing this statement with the earlier version of the Weitzenbock formula, and using 3.16, we see that we need only prove that

$$\sum_{i<j} c(e_i)c(e_j) R^S(e_i, e_j) = \tfrac{1}{4}\kappa.$$

Using the definition of R^S, the left hand side is equal to

$$\tfrac{1}{8} \sum_{i,j,k,l} R_{lkij} c(e_i e_j e_k e_l).$$

By 3.17 this equals

$$-\tfrac{1}{4} \sum_{a,l} \mathrm{Ric}_{la}\, c(e_a e_l);$$

and since the Ricci tensor is symmetric, all terms cancel here except those with $a = l$, which sum to $\tfrac{1}{4}\kappa$ as required. $\square$

The appearance of the scalar curvature in this context was first noted by Lichnerowicz [48].

Examples of Clifford bundles

EXAMPLE 3.19 THE EXTERIOR BUNDLE Let M be a Riemannian manifold; we use the metric to identify the bundles TM and T^*M. Let S denote the bundle $\bigwedge^* T^*M \otimes \mathbb{C}$. As a vector bundle this is naturally isomorphic to $\mathrm{Cl}(TM) \otimes \mathbb{C}$; the isomorphism simply converts a basis element $e_1 \wedge \ldots \wedge e_k$ ($e_1, \ldots, e_k$ orthonormal) for the exterior algebra into the basis element $e_1 \ldots e_k$ for the Clifford algebra. (Warning: This is not an isomorphism of *algebras*!) Using this isomorphism, the natural structure of $\mathrm{Cl}(TM) \otimes \mathbb{C}$ as a left module over itself can be transferred to $\bigwedge^* T^*M \otimes \mathbb{C}$, making this into a bundle of Clifford modules.

The Clifford action can be expressed concretely if we make use of the *interior product* operation in the exterior algebra; for a covector e, define for a k-form ω

$$e \lrcorner \omega = (-1)^{nk+n+1} * (e \wedge *\omega).$$

LEMMA 3.20 *The Clifford action of a covector e on $\omega \in \wedge^* T^*M$ is given by*

$$c(e)\omega = e \wedge \omega + e \lrcorner \omega .$$

PROOF Compute in an orthonormal basis. □

LEMMA 3.21 $\wedge^* T^*M \otimes \mathbb{C}$, *equipped with its natural metric and connection, is a Clifford bundle.*

PROOF We need to check the two properties listed in Definition (3.4). Recall that the metric on $\wedge^* T^*M$ can be defined in terms of the $*$-operation by $(\omega_1, \omega_2)\,\text{vol} = \omega_1 \wedge *\omega_2$.

Let ω_1 be a k-form, ω_2 a $(k-1)$-form. Then

$$\begin{aligned}
(\omega_2, e \lrcorner \omega_1)\,\text{vol} &= (-1)^{nk+n+1}\omega_2 \wedge * * (e \wedge *\omega_1) \\
&= (-1)^{nk+n+1+n(n-k+1)+(n-k+1)}\omega_2 \wedge e \wedge *\omega_1 \\
&= -(e \wedge \omega_2, \omega_1)\,\text{vol} .
\end{aligned}$$

This proves the skew-adjointness of the Clifford action. This calculation shows in fact that the interior multiplication is (up to a sign) the adjoint of exterior multiplication. The Levi-Civita connection is compatible with exterior multiplication and it is compatible with the metric, so it must also be compatible with interior multiplication, and this proves (3.4) ii). □

REMARK 3.22 As well as its left Clifford module structure, $\wedge^* T^*M \otimes \mathbb{C}$ also has a *right* module structure, coming from the right multiplication action of the Clifford algebra on itself; and these two structures commute, so that $\wedge^* T^*M$ is a *Clifford bimodule*. This bimodule structure will be important when we come to discuss the Witten complex of a manifold (9.14) and its applications to Morse theory.

(3.23) What is the Dirac operator of the Clifford bundle $\wedge^* T^*M \otimes \mathbb{C}$? Write

$$\begin{aligned}
D\omega &= \sum_i c(e_i) \nabla_i \omega \\
&= \sum_i e_i \wedge \nabla_i \omega + \sum_i e_i \lrcorner \nabla_i \omega \\
&= d\omega + d^*\omega
\end{aligned}$$

So D is the operator $d + d^*$ that we mentioned earlier, and $D^2 = dd^* + d^*d$ is the Laplacian. We call $D = d + d^*$ the *de Rham operator*.

EXAMPLE 3.24 Let V be any complex vector bundle equipped with a Hermitian metric and compatible connection. If S is a Clifford bundle then so is $S \otimes_\mathbb{C} V$. Its Dirac operator is often referred to as "the Dirac operator of S with coefficients in V".

EXAMPLE 3.25 COMPLEX MANIFOLDS The de Rham operator was constructed out of the regular representation of the algebra $\mathrm{Cl}(V) \otimes \mathbb{C}$; that is, we made the algebra act on itself by left multiplication. To obtain other interesting examples of Clifford bundles we need some more representations of the Clifford algebra. A fundamental example (called the *spin representation*, for reasons to be explained in the next chapter) arises when V is an even-dimensional (real) inner product space equipped with a *complex structure* — an operator $J \colon V \to V$ with $J^2 = -1$. One can choose J compatible with the metric: $(Jx, Jy) = (x, y)$. Then there is a decomposition

$$V \otimes \mathbb{C} = P \oplus Q$$

where P and Q are the $\pm i$ eigenspaces of $J \otimes 1$. They are a maximal transverse pair of isotropic subspaces — *isotropic* means that for $p_1, p_2 \in P, q_1, q_2 \in Q$ one has $(p_1, p_2) = 0 = (q_1, q_2)$. The inner product on $V \otimes \mathbb{C}$ places P and Q in duality.

Now we can make the exterior algebra $\Lambda^* P$ into a module over $\mathrm{Cl}(V) \otimes \mathbb{C}$ as follows. If $x \in \Lambda^* P$ and $p + q \in V \otimes \mathbb{C}$, with $p \in P$, $q \in Q$ define

$$(p + q).x = \sqrt{2}(p \wedge x + q \lrcorner x).$$

This extends to an action of the Clifford algebra since it satisfies the relations $p^2 = q^2 = 0$, $pq + qp = -2(p, q)$. So $\Lambda^* P$ becomes a representation of $\mathrm{Cl}(V) \otimes V$. Notice that if $\dim V = 2m$, then this representation has dimension 2^m; the regular representation, by contrast, has dimension 2^{2m}.

DEFINITION 3.26 The representation of $\mathrm{Cl}(V) \otimes \mathbb{C}$ defined in this way is called the *spin representation*.

Now let M be a $2m$-dimensional Riemannian manifold. The bundle $\mathrm{Cl}(TM) \otimes \mathbb{C}$ has its fiber at each point isomorphic to the Clifford algebra; but there may or may not exist a bundle S on M with fiber dimension 2^m such that, for each $x \in M$, the action of $\mathrm{Cl}(T_xM) \otimes \mathbb{C}$ on S_x is given by the spin representation. However, it is clear from the construction above that this will be the case when M is a Hermitian *complex* manifold. In this case each tangent space T_xM actually carries the structure of a complex vector space, and we can therefore define the operator J_x on T_xM simply as (complex) scalar multiplication by $\sqrt{-1}$. Applying our construction of the spin-representation to each fiber, we obtain a bundle S of Clifford modules. In fact, by construction $S = \bigwedge^* T_\mathbb{C} M \cong \bigwedge^* \overline{T}_\mathbb{C}^* M$, and via the usual constructions of complex geometry [36, Chapter 1], S acquires a Hermitian metric and connection which make it into a Clifford bundle in the sense of definition 3.4.

By construction, $C^\infty(S) \cong \bigoplus_q \Omega^{0,q}(M)$, in the usual (p,q)-decomposition of forms over the complex manifold M. Now the $(0,q)$-forms on a complex manifold form a cochain complex under the operator $\overline{\partial}$, called the *Dolbeault* complex:

$$\Omega^{0,0}(M) \xrightarrow{\overline{\partial}} \Omega^{0,1}(M) \xrightarrow{\overline{\partial}} \Omega^{0,2}(M) \to \cdots.$$

By analogy with the de Rham complex, one can ask about the relation of the Dirac operator D of S and the 'Dolbeault operator' $(\overline{\partial} + \overline{\partial}^*)$. We state the result without proof.

PROPOSITION 3.27 *If M is a Kähler manifold, then $D = \sqrt{2}(\overline{\partial} + \overline{\partial}^*)$. This identity does not hold for a general complex manifold, but the difference between the two sides is always an operator of 'zero order', that is an endomorphism of S.*

The extra complication is caused by the fact that there do not always exist "complex geodesic co-ordinates" on a complex manifold. The compatibility condition that allows such co-ordinates to exist is precisely the Kähler condition. See [36, page 107].

Notes

The factorization of a second order "Laplacian" operator into first order operator by means of the "Pauli spin matrices" (which are generators of a Clifford algebra) due to Dirac. Ziman ([74]) gives an account of the relevance of this factorization to quantum mechanics.

The concepts of a Clifford bundle and its associated Dirac operator are developed by Gromov and Lawson in [38]. A more detailed account, with many instructive examples, can be found in [47]. The discussion of the curvature of a Clifford bundle follows [12].

Exercises

QUESTION 3.28   Show that $\mathrm{Cl}(\mathbb{R}^1) \simeq \mathbb{C}, \mathrm{Cl}(\mathbb{R}^2) \simeq \mathbb{H}, \mathrm{Cl}(\mathbb{R}^3) \simeq \mathbb{H} \oplus \mathbb{H}$, where $\mathbb{H}$ denotes the quaternions and $\mathbb{R}^n$ is equipped with its usual positive definite form.

QUESTION 3.29   Let
$$x = \sum x_E . E$$
be an element of the Clifford algebra of $\mathbb{R}^{2m} \otimes \mathbb{C}$, where E runs over the standard basis of the Clifford algebra. Show that the trace of x as an endomorphism of the spin-representation is $2^m x_1$. (See Lemma 11.5.)

QUESTION 3.30   Prove Proposition 3.27. (See Gilkey [34], Section III.6.)

QUESTION 3.31   Compute the adjoint of the Clifford multiplication operator
$$c \colon C^\infty(TM \otimes S) \to C^\infty(S).$$
Use this and the formula for ∇^* of 3.9 to give another proof of the self-adjointness of the Dirac operator.

QUESTION 3.32   A *filtered algebra* A is an algebra which is written as the increasing union of subspaces $A_0, A_1, \ldots$, such that $A_i \cdot A_j \subseteq A_{i+j}$ for all i, j. The *associated graded algebra* $G(A)$ is the direct sum $\bigoplus_{i \geq 1} A_i / A_{i-1}$.

(i) Show that $G(A)$ inherits a well-defined multiplication from A.

(ii) Suppose that A is finite dimensional and equipped with an inner product. Show that there is a linear map $\sigma \colon A \to G(A)$ such that, if $a \in A_i \ominus A_{i-1}$, then $\sigma(a)$ coincides with the image of a in A_i / A_{i-1}.

(iii) Prove that σ is an isomorphism of vector spaces.

(iv) Let A be the Clifford algebra of a finite-dimensional vector space V equipped with a quadratic form. Show that A is a filtered algebra, if we define A_i to be the span of the products of i or fewer vectors of V. Show also that $G(A)$ is naturally isomorphic to the exterior algebra $\wedge^* V$.

This gives a more canonical approach to the vector space isomorphism between the Clifford and exterior algebras.

QUESTION 3.33 What is the 'Dirac operator' associated to $\wedge^* T^* M \otimes \mathbb{C}$ as a *right* Clifford module?

CHAPTER 4

The Spin groups

In this chapter we will study some important subgroups of the group of invertible elements in a Clifford algebra, and their representations. This material is needed to understand the geometrical significance of operators of Dirac type. However, it is not needed immediately in the development of the book, so the reader who is in a hurry to get on to some analysis can skip this chapter for now and come back to it later.

The Clifford algebra as a superalgebra

DEFINITION 4.1 An algebra A (over $\mathbb{R}$ or $\mathbb{C}$) is called a *superalgebra* (or $\mathbb{Z}/2$-graded algebra) if it is an internal direct sum of linear subspaces A_0 and A_1, with

$$A_0.A_0 \subset A_0, \quad A_1.A_1 \subset A_0, \quad A_0.A_1 \subset A_1, \quad A_1.A_0 \subset A_1.$$

The subspaces A_0 and A_1 are called the *even* and *odd* parts of the superalgebra A. The subset $A_0 \cup A_1$ is called the set of *homogeneous* elements of A, and if $x \in A_i$ is homogeneous, i is the *degree* of x, written $\deg(X)$; conventionally, $\deg(0) = 0$.

Equivalently, a superalgebra is an algebra A equipped with an automorphism ε, the *grading automorphism*, such that $\varepsilon^2 = 1$; the automorphism is defined by

$$\varepsilon(a_0 + a_1) = a_0 - a_1$$

$a_0 \in A_0, a_1 \in A_1$.

PROPOSITION 4.2 *The Clifford algebra of an orthogonal vector space V is a superalgebra, in which the elements of V are odd.*

PROOF Choose any basis for V, say (e_j), and define the even (resp. odd) part of the Clifford algebra to be the linear span of the basis elements $e_{j_1} \ldots e_{j_k}$, where k is even (resp. odd). The multiplication law for the Clifford algebra ensures that this gives a superalgebra structure independent of the choice of basis. □

If A is a superalgebra, then linear operators on A can be defined on homogeneous elements only, and extended by linearity to the whole of A. Thus we define the *super-commutator*
$$[x,y]_s = xy - (-1)^{\deg(x)\deg(y)} yx$$
on homogeneous elements; and the *super-center*
$$\mathfrak{Z}_s(A) = \{x \in A : [x,y]_s = 0 \quad \forall y \in A\}.$$

LEMMA 4.3 *If V is a real inner product space, then $\mathfrak{Z}_s(\mathrm{Cl}(V))$ is the scalar field $\mathbb{R}$, and $\mathfrak{Z}_s(\mathrm{Cl}(V) \otimes \mathbb{C})$ is the scalar field $\mathbb{C}$.*

PROOF Let $e_1 \ldots e_k$ be an orthonormal basis for V. Let $x \in \mathfrak{Z}_s(\mathrm{Cl}(V))$ and write
$$x = a + e_1 b$$
where a and b can be expanded in terms of basis elements that do not involve any e_1's. We may assume without loss of generality that x is homogeneous, so $\deg(x) = \deg(a) = \deg(b) + 1$. Now
$$\begin{aligned} xe_1 &= ae_1 + e_1 b e_1 \\ &= (-1)^{\deg(x)}(e_1 a - e_1^2 b) \\ &= (-1)^{\deg(x)}(e_1 a + b) \end{aligned}$$
but $e_1 x = e_1 a - b$.

Since $[x, e_1]_s = 0$, we deduce that $b = 0$. Thus x does not involve e_1. Similarly, it does not involve any other basis element, so it is a scalar. □

(4.4) Let an orientation for V be chosen. The *volume element* in $\mathrm{Cl}(V)$ is the product $\omega_V = e_1 \cdots e_k$, where $\{e_1, \ldots, e_k\}$ is a positively oriented orthonormal basis for V; one easily checks that the definition is independent of the choice of basis. Moreover, calculation shows that
$$\omega^2 = (-1)^{k(k+1)/2}, \qquad \omega v = (-1)^{k-1} v \omega \quad \forall v \in V.$$

Thus, if $k = 2m$ is even, the grading automorphism is an inner automorphism $\varepsilon(x) = \omega x \omega^{-1}$; moreover, $\omega^{-1} = (-1)^m \omega$. If $k = 2m+1$ is odd then ω is central in $\mathrm{Cl}(V)$ and $\omega^2 = (-1)^{m+1}$. In fact, it is easy to check that the center of $\mathrm{Cl}(V)$

consists just of the scalars if V is even-dimensional, and is spanned by 1 and ω if V is odd-dimensional. We leave this to the reader.

Groups of invertibles in the Clifford algebra

From now on, let $\mathrm{Cl}(k)$ denote the Clifford algebra of $\mathbb{R}^k$ with its usual positive definite form. In this section we will investigate certain subgroups of the group of invertible elements of $\mathrm{Cl}(k)$. Notice that, for $v \in \mathbb{R}^k$, $v \cdot v = -\|v\|^2$ in the Clifford algebra; so any nonzero v is an invertible in $\mathrm{Cl}(k)$.

DEFINITION 4.5

(i) The group $\mathrm{Pin}(k)$ is the multiplicative subgroup of $\mathrm{Cl}(k)$ generated by the unit vectors $v \in \mathbb{R}^n$.

(ii) The group $\mathrm{Spin}(k)$ is the even part of $\mathrm{Pin}(k)$, i.e. $\mathrm{Spin}(k) = \mathrm{Pin}(k) \cap \mathrm{Cl}(k)_0$.

Let $v \in \mathbb{R}^k$ be a unit vector. It is invertible in the Clifford algebra with inverse $v^{-1} = -v$. For $x \in V$ consider

$$-vxv^{-1} = vxv = x - 2(x,v)v$$

on expressing x in components parallel and perpendicular to the unit vector v. Notice that the right hand side of this equation can be described geometrically: it is the reflection of x in the hyperplane perpendicular to v. Since the unit vectors v generated the Pin group, we have proved that the *twisted adjoint representation* $\rho \colon \mathrm{Pin}(k) \to \mathrm{Aut}(\mathrm{Cl}(k))$ defined by

$$\rho(y)x = yx\varepsilon(y^{-1})$$

maps the subspace $\mathbb{R}^k$ of $\mathrm{Cl}(k)$ to itself by an orthogonal transformation (a product of reflections) and so gives a homomorphism $\rho \colon \mathrm{Pin}(k) \to O(k)$. Elements of $\mathrm{Spin}(k)$ are products of an even number of vectors in $\mathbb{R}^k$, so the restriction of ρ maps $\mathrm{Spin}(k) \to SO(k)$.

PROPOSITION 4.6 *There is an exact sequence*

$$0 \to \mathbb{Z}/2 \to \mathrm{Spin}(k) \xrightarrow{\rho} SO(k) \to 0$$

where $\mathbb{Z}/2 = \{\pm 1\} \subset \mathrm{Spin}(k)$.

PROOF It is well-known that every element of $SO(k)$ is a product of an even number of reflections, and in view of the calculation above this shows that ρ is surjective. An element y of the kernel of ρ must supercommute with every $v \in \mathbb{R}^k$; since such v generate the Clifford algebra, y must belong to $3_s(\mathrm{Cl}(k))$, and it must therefore be a scalar by 4.3. We need to show that the only scalars in $\mathrm{Spin}(k)$ are $\{\pm 1\}$.

For this purpose introduce the *transposition* antiautomorphism of $\mathrm{Cl}(k)$; if $x = v_1 \cdots v_m \in \mathrm{Cl}(k)$ is a product of basis vectors, we define $x^t = v_m \cdots v_1$. It is a simple exercise using the universal property of the Clifford algebra to show that $x \mapsto x^t$ is a well-defined antiautomorphism of $\mathrm{Cl}(k)$. But for a generator v of $\mathrm{Pin}(k)$ we clearly have $v^{-1} = v^t$; it follows that $x^{-1} = x^t$ for every $x \in \mathrm{Pin}(k)$. In particular if x is a scalar, $x^{-1} = x^t = x$, so x has square one. □

The exact sequence above displays $\mathrm{Spin}(k)$ as a double covering group of $SO(k)$; in particular, this shows that $\mathrm{Spin}(k)$ is a compact Lie group.

PROPOSITION 4.7 *For $k \geq 2$, the group $\mathrm{Spin}(k)$ is connected; for $k \geq 3$ it is simply connected, and the exact sequence above displays $\mathrm{Spin}(k)$ as the universal cover of $SO(k)$.*

PROOF Consider a part of the exact homotopy sequence of the fibration

$$0 \to \mathbb{Z}/2 \to \mathrm{Spin}(k) \to SO(k) \to 0.$$

The exact homotopy sequence gives

$$\pi_1 \mathbb{Z}/2 \to \pi_1 \mathrm{Spin}(k) \to \pi_1 SO(k) \to \pi_0 \mathbb{Z}/2 \to \pi_0 \mathrm{Spin}(k) \to \pi_0 SO(k)$$

where we know $\pi_1 \mathbb{Z}/2$ and $\pi_0 SO(k)$ are trivial, $\pi_0 \mathbb{Z}/2$ is $\mathbb{Z}/2$, and $\pi_1 SO(k)$ is $\mathbb{Z}/2$ if $k \geq 3$. It is enough to show that the map $\pi_0 \mathbb{Z}/2 \to \pi_0 \mathrm{Spin}(k)$ is trivial. This amounts to showing that the points $+1$ and -1 are connected in $\mathrm{Spin}(k)$, which they are by the path

$$t \mapsto \cos t + e_1 e_2 \sin t$$

provided that $k \geq 2$. □

Since $\rho : \mathrm{Spin}(k) \to SO(k)$ is a covering map, there is a natural identification of the Lie algebra of $\mathrm{Spin}(k)$ with that of $SO(k)$, which is the Lie algebra of antisymmetric $k \times k$ matrices. On the other hand, since $\mathrm{Spin}(k)$ is a submanifold of the vector space $\mathrm{Cl}(k)$, its Lie algebra can be identified with a vector subspace of $\mathrm{Cl}(k)$. What is the relationship between these identifications?

LEMMA 4.8 *The Lie algebra of* $\mathrm{Spin}(k)$ *may be identified with the vector subspace of* $\mathrm{Cl}(k)$ *spanned by the products* $e_i e_j, i \neq j$. *The identification associates an antisymmetric matrix* a_{ij} *with the element* $\frac{1}{4} \sum_{i,j} a_{ij} e_i e_j$ *of* $\mathrm{Cl}(k)$.

PROOF Since $(e_i e_j)^2 = -1$,

$$\exp(t e_i e_j) = \cos t + e_i e_j \sin t \in \mathrm{Spin}(k).$$

Thus all the $e_i e_j$ belong to the Lie algebra; and they span it, since it has dimension $\frac{1}{2}k(k-1)$.

Now

$$\rho(x)v = xvx^{-1} \quad (\text{for } x \in \mathrm{Spin}(k)) = \mathrm{Ad}(x)v;$$

so if u belongs to the Lie algebra, $\rho_*(u)v = \mathrm{ad}(u)v = [u, v]$. If $u = e_1 e_2$ (say), we compute

$$\mathrm{ad}(u)e_1 = 2e_2, \qquad \mathrm{ad}(u)e_2 = -2e_1, \qquad \mathrm{ad}(u)e_i = 0 \quad (i \neq 1, 2).$$

So $\mathrm{ad}(u)$ is represented by the matrix

$$a_{ij} = 2(\delta_{i1}\delta_{j2} - \delta_{i2}\delta_{j1});$$

thus, $\sum a_{ij} e_i e_j = 2(e_1 e_2 - e_2 e_1) = 4u$. The result follows. □

Representation theory of the Clifford algebra

We now want to study the representations of the Spin group and of the Clifford algebra. We will do this by looking at the representation theory of a finite multiplicative subgroup. Let $e_1, \ldots, e_k$ be the standard orthonormal basis of $\mathbb{R}^k$ and let $\Gamma_k \subseteq \mathrm{Pin}(k)$ be the group of order 2^{k+1} consisting of all the elements

$$\pm e_1^{i_1} e_2^{i_2} \cdots e_k^{i_k}$$

59

where each of $i_1, i_2, \ldots, i_k$ is either 0 or 1. In particular, E_k contains $(-1) \in \mathrm{Cl}(k)$; denote this by ν when it is considered as an element of E_k.

PROPOSITION 4.9 *There is a 1-1 correspondence between*

(i) *Representations of the Clifford algebra* $\mathrm{Cl}(k)$;
(ii) *Representations of* $\mathrm{Pin}(k)$ *on which* ν *acts as* -1;
(iii) *Representations of* E_k *on which* ν *acts as* -1.

The proof is obvious. In the remainder of this section we will take 'representation' to mean 'complex representation', although the proposition is clearly valid over the real field as well.

By the proposition, one can use the complex representation theory of the finite group E_k to study that of $\mathrm{Cl}(k) \otimes \mathbb{C}$. We will assume that the reader is familiar with the representation theory of finite groups, and in particular with the properties of group characters; a possible reference for this subject is [46].

Since ν is a central involution in the finite group E_k, it must act as $+1$ or -1 on each irreducible representation of E. Those irreducible representations on which ν acts as $+1$ are representations of the abelian group $E_k/(\nu)$ of order 2^k, so there are 2^k of them. How many more representations does E_k have? Here a distinction makes itself apparent according to whether k is even or odd.

LEMMA 4.10

(a) *If k is even, the center of E_k is* $\{1, \nu\}$.
(b) *If k is odd, the center of E_k is* $\{1, \nu, \omega, \nu\omega\}$, *where ω is the volume element* (4.4) *in the Clifford algebra*.

PROOF Let $g = e_1^{i_1} \cdots e_k^{i_k}$. If $i_r = 1, i_s = 0$ then one can check by hand that $e_r e_s g = \nu g e_r e_s$. So the only possible central elements are $1, \nu, \omega = e_1 \cdots e_k$, and $\nu\omega$. If k is odd, we have seen that ω is central; if k is even

$$e_1 \omega = \nu \omega e_1$$

so ω is noncentral. The result follows. □

Now we count conjugacy classes in E_k. The conjugacy class of $g \in E_k$ must be either $\{g\}$ (if g is central) or $\{g, \nu g\}$ (otherwise); this is an easy consequence of the fact that $E/\langle\nu\rangle$ is abelian. The number of conjugacy classes is therefore

$$\frac{2^{k+1} - 2}{2} + 2 = 2^k + 1 \ (k \text{ even}), \qquad \frac{2^{k+1} - 4}{2} + 4 = 2^k + 2 \ (k \text{ odd}).$$

We have seen that E_k has 2^k irreducible representations on which ν acts as $+1$. Since the number of conjugacy classes is equal to the number of irreducible representations, it follows that if k is even E_k has just one more irreducible representation, on which ν must act as -1; and that if k is odd E_k has two inequivalent irreducible representations on which ν acts as -1.

One must consider the two cases separately from now on, and we will concentrate on the even case, where $k = 2m$. (The odd case is considered in the exercises.) Our argument shows that $\mathrm{Cl}(k) \otimes \mathbb{C}$ has just one irreducible representation, which is denoted Δ and called the *spin representation*. Its dimension can be calculated if we recall that the sum of the squares of the dimensions of the representations of E_k is equal to the order of E_k, so

$$2^k + (\dim \Delta)^2 = 2^{n+1}.$$

Therefore, $\dim \Delta = 2^m$.

Since Δ is its only irreducible representation, $\mathrm{Cl}(k) \otimes \mathbb{C}$ is isomorphic to the matrix algebra $\mathrm{End}(\Delta)$. As a check on this, note that

$$\dim(\mathrm{End}(\Delta)) = (2^m)^2 = 2^k = \dim(\mathrm{Cl}(k) \otimes \mathbb{C}).$$

In 3.26 we gave a concrete construction of a representation of $\mathrm{Cl}(k) \otimes \mathbb{C}$, of dimension 2^m. By dimension counting, this must in fact be (isomorphic to) the spin representation.

REMARK 4.11 As a complex representation of the finite group E_k, Δ is provided with a Hermitian metric with respect to which E_k acts by unitary transformations. Since each generator e_i has square -1 and is unitary on Δ, it must in fact be skew-adjoint on Δ. So the action of the Clifford algebra verifies the first part of definition 3.4.

(4.12) Any finite-dimensional complex representation W of $\mathrm{Cl}(k)$ must be a direct sum of copies of Δ, or to put it another way $W = \Delta \otimes V$ for some auxiliary 'coefficient' vector space V. Notice that V can be recovered from W as $\mathrm{Hom}_{\mathrm{Cl}(k)}(\Delta, W)$. Moreover, we have

$$\mathrm{End}_{\mathbb{C}}(W) = \mathrm{Cl}(k) \otimes \mathrm{End}_{\mathbb{C}}(V) = \mathrm{Cl}(k) \otimes \mathrm{End}_{\mathrm{Cl}(k) \otimes \mathbb{C}}(W).$$

DEFINITION 4.13 Let F be a Clifford module endomorphism of a representation $W = \Delta \otimes V$ of $\mathrm{Cl}(k)$. Its *relative trace* $\mathrm{tr}^{W/\Delta}(F)$ is defined to be the trace of the $\mathbb{C}$-linear endomorphism of V corresponding to F under the identification $\mathrm{End}_{\mathrm{Cl}(k) \otimes \mathbb{C}}(W) = \mathrm{End}_{\mathbb{C}}(V)$.

(4.14) Notice that Δ is also an irreducible representation of $\mathrm{Pin}(k)$. Representation theory allows two possibilities for the restriction of Δ to the index two normal subgroup $\mathrm{Spin}(k)$; either the restriction is irreducible, or it splits as the direct sum of two inequivalent irreducible representations of the same dimension. To see that the latter case actually occurs, recall that the volume element $\omega \in \mathrm{Cl}(k)$ has $\omega^2 = (-1)^m$, and that $\omega x = \varepsilon(x)\omega$ for all $x \in \mathrm{Cl}(k)$. Let Δ_+ and Δ_- be the ± 1 eigenspaces of $i^m \omega$ acting on Δ; then $\mathrm{Cl}(k) \otimes \mathbb{C}$ acts on $\Delta = \Delta_+ \oplus \Delta_-$ in such a way that even elements of the Clifford algebra preserve this direct sum decomposition, and odd elements reverse it. In particular, Δ^+ and Δ^- are themselves representations of $\mathrm{Spin}(k)$; they are called the positive and negative *half-spin representations*, and they are irreducible.

REMARK 4.15 We can reformulate this by saying that the super vector space $\Delta = \Delta_+ \oplus \Delta_-$ becomes a *graded representation* of $\mathrm{Cl}(k)$.

Spin structures on manifolds

Now let M be an oriented Riemannian manifold, of dimension n, and let E be the principal $SO(n)$-bundle of oriented orthonormal frames for the tangent bundle. We will assume n is even, but an analogous discussion can be made for odd n, once the theory of the spin representation has been worked out.

DEFINITION 4.16 A *spin-structure* on M is a principal $\mathrm{Spin}(n)$-bundle $\tilde{E}$ over M which is a double covering of E such that the restriction to each fiber of the double covering $\tilde{E} \to E$ is the double covering $\rho\colon \mathrm{Spin}(n) \to SO(n)$. If M admits a spin structure, it is called a *spin manifold*.

We don't want to go too deeply into questions about the existence and uniqueness of spin structures here. However, we should at least prove that spin structures are not uncommon:

PROPOSITION 4.17 *If M is 2-connected, then it admits a unique spin structure.*

PROOF The double coverings of the connected space E are classified by the homomorphisms of the fundamental group $\pi_1 E$ to $\mathbb{Z}/2$; and those double coverings that restrict on a fiber to the standard double covering of $SO(n)$ are those classified by homomorphisms $\pi_1 E \to \mathbb{Z}/2$ such that the composite $\pi_1 SO(n) \to \pi_1 E \to \mathbb{Z}/2$ is an isomorphism. But if M is 2-connected, the exact homotopy sequence gives
$$0 = \pi_2 M \to \pi_1 SO(n) \to \pi_1 E \to \pi_1 M = 0$$
so $\pi_1 SO(n) \to \pi_1 E$ is an isomorphism, and M has a unique spin structure. □

DEFINITION 4.18 If M is a spin manifold, then its *spin bundle* Δ is the vector bundle associated to the principal spin bundle by means of the spin representation.

DEFINITION 4.19 The *spin connection* on the principal $\mathrm{Spin}(n)$ bundle $\tilde{E}$ over a spin manifold M is defined to be the lifting to $\tilde{E}$ of the principal $SO(n)$ connection on E induced by the Levi-Civita connection on TM. The *spin connection* on Δ is the connection on Δ associated (via the spin representation) to the spin connection on $\tilde{E}$.

Since the spin representation is unitary, the bundle Δ has a natural hermitian metric. Moreover, the spin connection is compatible with this metric. Thus we conclude

PROPOSITION 4.20 *The spin bundle Δ (equipped with its Hermitian metric and spin connection) over a spin manifold M is a Clifford bundle (3.4).*

The fundamental nature of the spin representation is revealed by

PROPOSITION 4.21 *The twisting curvature (3.16) of the spin bundle associated to a spin structure is zero.*

PROOF Let $\{e_k\}$ be a local orthonormal frame for TM. Recall that the connection and curvature forms for TM have their values in the Lie algebra $\mathfrak{so}(n)$ of antisymmetric matrices. In particular, the curvature is an $\mathfrak{so}(n)$-valued two-form, whose matrix entries are (Re_k, e_l) where R denotes the Riemann curvature operator. By lemma 4.8, the corresponding $\mathfrak{spin}(n)$-valued two-form (which gives the curvature of the spin connection) is

$$\tfrac{1}{4}\sum_{k,l}(Re_k, e_l)e_k e_l$$

and this acts on the spin representation by

$$\tfrac{1}{4}\sum_{k,l}(Re_k, e_l)c(e_k)c(e_l)$$

which is exactly the $\text{End}(\Delta)$-valued 2-form R^Δ of definition 3.14. The result now follows from the definition of twisting curvature (3.16). □

REMARK 4.22 Suppose now that S is *any* Clifford bundle over a spin manifold M. Then there is a vector bundle $V = \text{Hom}_{\text{Cl}}(\Delta, S)$, equipped with hermitian metric and connection, such that $S \cong \Delta \otimes V$ as Clifford bundles. The curvature of the natural connection on a tensor product of this type is (in an obvious notation)

$$K^\Delta \otimes 1 + 1 \otimes K^V.$$

Proposition 4.21 identifies the first term as the Riemann endomorphism (3.14) of S and the second as the twisting curvature (3.16).

Spin bundles and characteristic classes

Let M be a spin manifold, of even dimension $2m$, and let Δ the associated spin bundle. We will need to know the Chern character of the complex vector bundle Δ.

PROPOSITION 4.23 *The Chern character* $\text{ch}(\Delta)$ *of the spin bundle is equal to* $2^m \mathfrak{S}(TM)$, *where* $\mathfrak{S}$ *denotes the Pontrjagin genus associated to the holomorphic function* $g(z) = \cosh(\tfrac{1}{2}\sqrt{z})$.

We will denote the Pontrjagin genus associated to this particular holomorphic function by $V \mapsto \mathfrak{G}(V)$.

PROOF In general, an even-dimensional oriented Euclidean (real) vector bundle V is called a *spin vector bundle* if there is given a double covering of its associated SO principal bundle of oriented orthonormal frames satisfying the conditions of (4.16). (Thus, the tangent bundle to a spin manifold is a spin bundle.) Given such a spin vector bundle we can form its spin bundle $\Delta(V)$ as above and we want to show that

$$\mathrm{ch}(\Delta(V)) = 2^{\frac{1}{2}\dim V}\mathfrak{G}(V) \tag{4.24}$$

We regard this as a pointwise identity between certain polynomial functions of the curvature of V. To prove it, we may therefore assume that the curvature is block diagonal, so that V is a direct sum of 2-dimensional bundles. Moreover, if $V = V_1 \oplus V_2$ (with V_1, V_2 even-dimensional) then $\Delta(V) = \Delta(V_1) \otimes \Delta(V_2)$, so both sides of (4.24) are multiplicative on direct sums; thus it suffices to consider 2-dimensional V. Such a V can also be regarded as a 1-dimensional complex vector bundle, then denoted V_c. On the other hand, $\Delta(V)$ is a 2-dimensional complex vector bundle, decomposed by the grading operator into the direct sum of two 1-dimensional complex components $\Delta^+(V)$ and $\Delta^-(V)$. I claim that

$$\Delta^+ \otimes_{\mathbb{C}} \Delta^+ \cong V_c, \qquad \Delta^- \otimes_{\mathbb{C}} \Delta^- \cong V_c^*$$

as vector bundles, or equivalently as representations of Spin(2). This must be checked explicitly.

The Clifford algebra $\mathrm{Cl}(\mathbb{R}^2) \otimes \mathbb{C}$ is isomorphic to $M_2(\mathbb{C})$ and it is spanned by the four matrices

$$1 = \begin{pmatrix} 1 & 0 \\ 0 & 1 \end{pmatrix}, \quad e_1 = \begin{pmatrix} i & 0 \\ 0 & -i \end{pmatrix}, \quad e_2 = \begin{pmatrix} 0 & i \\ i & 0 \end{pmatrix}, \quad e_1 e_2 = \begin{pmatrix} 0 & -1 \\ 1 & 0 \end{pmatrix}.$$

It is not hard to verify directly from the definition that, in this representation, Spin(2) is the rotation group consisting of matrices

$$\begin{pmatrix} \cos\theta & -\sin\theta \\ \sin\theta & \cos\theta \end{pmatrix}$$

and that the action of an element of Spin(2) on an element of $\mathbb{R}^2$ represented by the matrix
$$i \begin{pmatrix} x & y \\ y & -x \end{pmatrix}$$
rotates it through 2θ, as we should expect.

On the other hand, the grading operator is $ie_1 e_2$, which equals
$$\begin{pmatrix} 0 & -i \\ i & 0 \end{pmatrix}$$
so the eigenspaces Δ^+ and Δ^- are given by
$$\Delta^+ = \left\{ \begin{pmatrix} x \\ -ix \end{pmatrix} : x \in \mathbb{C} \right\}, \quad \Delta^- = \left\{ \begin{pmatrix} x \\ ix \end{pmatrix} : x \in \mathbb{C} \right\}.$$
The action of an element of Spin(2) on Δ^+ is given by
$$\begin{pmatrix} \cos\theta & -\sin\theta \\ \sin\theta & \cos\theta \end{pmatrix} \begin{pmatrix} x \\ -ix \end{pmatrix} = \begin{pmatrix} (\cos\theta + i\sin\theta)x \\ -i(\cos\theta + i\sin\theta)x \end{pmatrix};$$
that is rotation through θ. So $\Delta^+ \otimes \Delta^+ \cong V_c$, and similarly $\Delta^- \otimes \Delta^- \cong V_c^*$.

Now let x denote the first Chern class $c_1(\Delta^+)$. Then
$$\operatorname{ch}(\Delta) = \operatorname{ch}(\Delta^+) + \operatorname{ch}(\Delta^-) = e^x + e^{-x} = 2\cosh x.$$

On the other hand,
$$p_1(V) = -c_2(V \otimes \mathbb{C}) = -c_2(V_c \oplus V_c^*) = -c_1(V_c)c_1(V_c^*) = 4x^2.$$

The result now follows from the definition of the Pontrjagin genus $\mathfrak{S}$. □

(4.25) Now let S be a general Clifford bundle on M. By definition, the *relative Chern character* of S is the cohomology class represented by the differential form
$$\operatorname{ch}(S/\Delta) = \operatorname{tr}^{S/\Delta}(\exp(-F^S/2\pi i))$$
where F^S is the twisting curvature and $\operatorname{tr}^{S/\Delta}$ is the relative trace of 4.13. Notice that, if $S = \Delta \otimes V$, then $\operatorname{ch}(S/\Delta)$ is just the ordinary Chern character of V. (The fact that such a decomposition is always possible locally shows that $\operatorname{ch}(S/\Delta)$ is indeed a

closed form, and so does represent a cohomology class.) From the above results we have

$$\operatorname{ch}(S) = 2^m \mathfrak{S}(TM) \operatorname{ch}(S/\Delta).$$

The complex Spin group

DEFINITION 4.26 The group $\operatorname{Spin}^c(k)$ is the subgroup of $\operatorname{Cl}(k) \otimes \mathbb{C}$ generated by $\operatorname{Spin}(k)$ together with the circle S^1 of unit complex numbers.

Notice that S^1 belongs to the center of $\operatorname{Cl}(k) \otimes \mathbb{C}$; thus we have an epimorphism $\operatorname{Spin}(k) \times S^1 \to \operatorname{Spin}^c(k)$. The kernel of this epimorphism consists of pairs (λ^{-1}, λ), where $\lambda \in S^1 \cap \operatorname{Spin}(k)$. But we have already remarked that the only scalars in $\operatorname{Spin}(k)$ are ± 1, so we get the isomorphism

$$\operatorname{Spin}^c(k) \cong \operatorname{Spin}(k) \times_{\{\pm 1\}} S^1$$

where the notation refers to the quotient of the product $\operatorname{Spin}(k) \times S^1$ by $\{(\lambda^{-1}, \lambda) : \lambda \in \{\pm 1\}\}$.

If k is even, it is clear that the spin representation (and also the half-spin representations) are representations of the group Spin^c as well as of Spin; since they are in fact representations of the complexified Clifford algebra.

PROPOSITION 4.27 *There is a short exact sequence*

$$0 \to \mathbb{Z}/2 \to \operatorname{Spin}^c(k) \to SO(k) \times S^1 \to 1$$

where the $\mathbb{Z}/2$ subgroup is generated by $[(-1, 1)] = [(1, -1)]$.

The map $\operatorname{Spin}^c(k) \to SO(k) \times S^1$ is $(x, \lambda) \mapsto (\rho(x), \lambda^2)$, where $\rho\colon \operatorname{Spin}(k) \to SO(k)$ is the double covering. Notice that the composite

$$S^1 \to \operatorname{Spin}^c(k) \to S^1$$

is the double covering map, not the identity.

Let M be an oriented Riemannian manifold. Let E denote the oriented orthonormal frame bundle of M, as before, and let L be a principal S^1-bundle on M (the obvious representation of S^1 on $\mathbb{C}$ allows us to regard L as a complex hermitian line bundle, whence the notation).

DEFINITION 4.28 A *Spin^c-structure* on M is a principal Spin$^c(n)$-bundle $\tilde{E}$ over M which is a double covering of $E \times L$ such that the restriction to each fiber of the double covering $\tilde{E} \to E$ is the double covering $\rho : \text{Spin}^c(n) \to SO(n) \times S^1$. The bundle L is called the *fundamental line bundle* associated to the Spin^c-structure.

Using the spin representation, as before, we can associate a spin bundle S to any Spin^c-structure. It has a hermitian metric; to equip it with a compatible connection we must choose a connection on L, and then lift the product $SO(n) \times S^1$ connection on $E \times L$ to a Spin$^c(n)$ connection on $\tilde{E}$. The analogue of proposition 4.21 is

PROPOSITION 4.29 *The twisting curvature of the spin bundle associated to a Spin^c structure is $\frac{1}{2}F$, where F is the curvature operator of the chosen connection on the fundamental line bundle L.*

The proof is left to the reader.

REMARK 4.30 Let M be a Riemannian manifold and suppose that there exists a Clifford bundle S over M whose fiber at each point is a copy of the spin representation. Then it can be shown (see exercise 4.36) that M admits a Spin^c structure for which S is the associated spin bundle. In particular, the fundamental line bundle L can be recovered from S; it is simply the bundle $\text{Hom}_{\text{Cl} \otimes \mathbb{C}}(\bar{S}, S)$ of module-isomorphisms between the representation S and its complex conjugate.

Notes

Fundamental references on Clifford modules are the paper [4] and the book [47], both of which give a much more systematic development than we have done. The approach taken here, by way of the finite groups E_k, comes from unpublished lecture notes of the late J.F. Adams.

Exercises

QUESTION 4.31 Using the fibration $SO(n-1) \to SO(n) \to S^{n-1}$, verify the assertion in the text that $\pi_1 SO(n) \cong \mathbb{Z}/2$ for $n \geq 3$.

QUESTION 4.32 Show that the even part of $Cl(k)$ is isomorphic to $Cl(k-1)$. In particular, identify the even part of $Cl(3) \otimes \mathbb{C}$ with a matrix algebra $M_2(\mathbb{C})$. Using this identification, construct an isomorphism $\text{Spin}(3) \cong SU(2)$.

QUESTION 4.33 The graded tensor product of two super vector spaces $U = U_+ \oplus U_-$ and $V = V_+ \oplus V_-$ is their ordinary tensor product as vector spaces, with grading defined by
$$(U \hat{\otimes} V)_+ = U_+ \otimes V_+ + U_- \otimes V_-, \quad (U \hat{\otimes} V)_- = U_+ \otimes V_- + U_- \otimes V_+.$$
Prove the graded analogue of 4.12, namely that the most general graded representation of $Cl(k)$ for k even is of the form $\Delta \hat{\otimes} V$ for some graded vector space V.

QUESTION 4.34 Suppose that we consider the spin bundle Δ of an even-dimensional spin manifold as a *super* vector bundle by way of the decomposition $\Delta = \Delta^+ \oplus \Delta^-$. (Notice that this grading depends on the orientation of M.) Prove that the super Chern character $\text{ch}_s(\Delta)$ is equal to $e(TM)$, where $e(TM)$ denotes the Euler class of question 2.36.

QUESTION 4.35 Work out (using the method in the text) the complex representation theory of the Clifford algebra of an *odd*-dimensional Euclidean space.

QUESTION 4.36 Let $\alpha: \text{Spin}^c(2k) \to U(2^k)$ be the homomorphism arising from the spin representation.

(i) Show that there is a pull-back diagram

$$\begin{array}{ccc} \text{Spin}^c(2k) & \xrightarrow{\alpha} & U(2^k) \\ \downarrow & & \downarrow \pi \\ SO(2k) & \xrightarrow{\varphi} & PU(2^k) \end{array}$$

where $PU(2^k) = U(2^k)/U(1)$ is the projective unitary group, π is the obvious quotient map, and φ is the projectivization of the spin representation.

(ii) Suppose that M is a $2k$-dimensional manifold which admits a Clifford bundle S whose fiber at each point is a copy of the spin representation, so that $Cl(TM) \otimes \mathbb{C} \cong \text{End}(S)$. Let E be the orthonormal frame bundle of M, and

E' be the complex orthonormal frame bundle of S. Show that $\pi_* E \cong \varphi_* E'$ as principal $PU(2^k)$-bundles. Deduce that there is a principal $\text{Spin}^c(2k)$ bundle E'' over M which covers E. Thus M is a Spin^c manifold. See Plymen [59].

QUESTION 4.37 In this question we will consider the general classification of spin structures; some information from homotopy theory will be required. Let M be a compact oriented Riemannian manifold, and let E be its principal $SO(n)$ frame bundle (n even).

(a) By considering the Serre spectral sequence of the fibration $SO(n) \to E \to M$, derive the exact sequence
$$0 \to H^1(M; \mathbb{Z}_2) \xrightarrow{\pi^*} H^1(E; \mathbb{Z}_2) \xrightarrow{i^*} H^1(SO(n); \mathbb{Z}_2) \xrightarrow{\delta} H^2(M; \mathbb{Z}_2).$$

(b) Show that the set of spin structures on M may be identified with the complement of $\text{Ker}(i^*)$ in $H^1(E; \mathbb{Z}_2)$.

(c) The image under δ of the generator of $H^1(SO(n); \mathbb{Z}_2)$ is a characteristic class of M, called the *second Stiefel-Whitney class* $w_2(M)$. Show that M admits a spin structure iff $w_2(M) = 0$, and that if this is so, the number of distinct spin structures is equal to the number of elements in $H^1(M; \mathbb{Z}_2)$.

- It can be shown [55, page 171] that the Stiefel-Whitney classes of a complex manifold are the mod 2 reductions of its Chern classes. Use this fact to show that $\mathbb{CP}^r$ has no spin structure.

CHAPTER 5

Analytic properties of Dirac operators

A *harmonic function* u (for instance on $\mathbb{R}^2$) is a solution of the Laplace equation
$$\frac{\partial^2 u}{\partial x^2} + \frac{\partial^2 u}{\partial y^2} = 0 .$$
One of the basic properties of harmonic functions is that they are smoother than you think; though u need only be twice differentiable for the equation above to make sense, one knows from complex variable theory that u is locally the real part of a holomorphic function, and hence is infinitely differentiable.

It turns out that the more general Dirac operators we have been considering have analogous properties. To obtain these properties we need a quantitative measure of the degree of differentiability of a function on a compact manifold. Such a measure is provided by the Sobolev spaces which we will now study.

Sobolev Spaces

These are defined by Fourier series. Initially, therefore, we work on the torus
$$\mathbb{T}^n = \mathbb{R}^n / 2\pi \mathbb{Z}^n .$$

DEFINITION 5.1 Let $f \colon \mathbb{T}^n \to \mathbb{C}$ be an integrable function. The *Fourier series* for f is the formal series
$$\sum_{\nu \in \mathbb{Z}^n} a_\nu e^{i\nu . x}$$
where
$$a_\nu = \hat{f}(\nu) = \frac{1}{(2\pi)^n} \int_{\mathbb{T}^n} f(x) e^{-i\nu . x} \, dx .$$

When f is a trigonometric polynomial it is equal to its own Fourier series. Many delicate results describe conditions under which the Fourier series converges to f, but for our purposes some of the simplest will suffice, which all follow from the fact that the functions $e_\nu \colon x \mapsto (2\pi)^{-n/2} e^{i\nu . x}$ form an orthonormal basis of the Hilbert space $L^2(\mathbb{T}^n)$.

PARSEVAL'S THEOREM 5.2 For $f \in L^2(\mathbb{T}^n)$,

$$\int_{\mathbb{T}^n} |f|^2 = (2\pi)^n \sum_\nu |\hat{f}(\nu)|^2 .$$

INVERSION THEOREM FOR L^2 5.3 For $f \in L^2(\mathbb{T}^n)$, the Fourier series of f converges in the L^2-norm to f.

INVERSION THEOREM FOR C^∞ 5.4 For $f \in C^\infty(\mathbb{T}^n)$, the Fourier series of f converges in the Fréchet C^∞ topology to f. The Fourier coefficients $\hat{f}(\nu)$ are *rapidly decreasing*: for any k there is a constant C_k such that $|\hat{f}(\nu)| \leqslant C_k(1+|\nu|)^{-k}$.

(The Fréchet topology of $C^\infty(\mathbb{T}^n)$ is the topology of uniform convergence of all derivatives.)

The Fourier transform converts differentiation into multiplication. Thus, statements about the differentiability of a function f on $\mathbb{T}^n$ may be translated into statements about the rate of growth of its Fourier coefficients.

DEFINITION 5.5 Let k be a positive integer. The *Sobolev k-inner product* on $C^\infty(\mathbb{T}^n)$ is defined by the formula

$$\langle f_1, f_2 \rangle_k = (2\pi)^n \sum_\nu \hat{f}_1(\nu)\overline{\hat{f}_2(\nu)}(1+|\nu|^2)^k .$$

(This makes sense since $\hat{f}_1$ and $\hat{f}_2$ are rapidly decreasing.)

The *Sobolev k-norm* is the norm induced by this inner product. The k'th *Sobolev space*, denoted W^k, is the completion of $C^\infty(\mathbb{T}^n)$ in the k-norm.

By Parseval's theorem, W^0 is isometrically isomorphic to L^2. The space W^k can be thought of as the space of functions whose first k derivatives belong to L^2; making this statement precise, however, requires some distribution theory.

There are three basic facts about Sobolev spaces, given in the next three propositions.

PROPOSITION 5.6 *The space $C^k(\mathbb{T}^n)$ of k times continuously differentiable functions is a subspace of $W^k(\mathbb{T}^n)$, and the inclusion map is continuous.*

PROOF If $f \in C^k(\mathbb{T}^n)$, then one can differentiate the Fourier series formally to obtain

$$\hat{f}(\nu) = \left(\frac{1}{i\nu_j}\right)^k \hat{f}_j(\nu)$$

where $f_j = (\partial/\partial x^j)^k f$. (To prove this integrate by parts k times in the formula for $\hat{f}$.) Each f_j is continuous, hence square integrable, so $\hat{f}_j$ belongs to l^2 by Parseval's theorem. Therefore

$$\nu \mapsto \hat{f}(\nu)(1+|\nu|)^k$$

belongs to ℓ^2, which means exactly that $f \in W^k$. The asserted continuity is easy to check, either directly or by means of the closed graph theorem. □

SOBOLEV EMBEDDING THEOREM: 5.7 For any integer $p > n/2$, the space W^{k+p} is continuously included in C^k.

PROOF Let $f \in W^{k+p}$; then

$$\sum_\nu |\hat{f}(\nu)|^2 (1+|\nu|^2)^k (1+|\nu|^2)^p < \infty .$$

By Cauchy-Schwartz, then,

$$\left(\sum_\nu |\hat{f}(\nu)|(1+|\nu|^2)^{k/2}\right)^2 \leq \left(\sum_\nu |\hat{f}(\nu)|^2(1+|\nu|^2)^k(1+|\nu|^2)^p\right)\left(\sum_\nu (1+|\nu|^2)^{-p}\right)$$

and this is finite since $p > n/2$. Therefore, $\sum_\nu |\nu|^k |\hat{f}(\nu)| < \infty$, so that the Fourier series for the first k derivatives of f converges absolutely and uniformly. □

RELLICH'S THEOREM: 5.8 If $k_1 < k_2$ then the inclusion operator $W^{k_2} \to W^{k_1}$ is a compact linear operator.

PROOF Let B be the unit ball of W^{k_2}. Given $\varepsilon > 0$, one can choose a subspace Z of W^{k_2} of finite codimension with the property that for all $f \in B \cap Z$, $\|f\|_{k_1} < \varepsilon$: just take Z to be the space $\{f : \hat{f}(\nu) = 0 \text{ for } |\nu| < N\}$, for a suitably large N. The unit ball of W^{k_2}/Z is precompact, so can be covered by finitely many balls of radius ε. Hence B can be covered by finitely many balls of radius 2ε in the W^{k_1} norm. Since ε is arbitrary, B is precompact in that norm. □

Now we will define Sobolev spaces on manifolds other than $\mathbb{T}^n$. To do this we need to give a different definition of the Sobolev norms.

PROPOSITION 5.9 *The Sobolev k-norm on $C^\infty(\mathbb{T}^n)$ is equivalent to the norm given by*

$$f \mapsto \sum_{|\alpha| \leqslant k} \|\frac{\partial f}{\partial x^\alpha}\|$$

where the norms on the right are L^2 norms.

PROOF This is a straightforward application of the argument used in proving proposition (5.6). □

COROLLARY 5.10 *Multiplication by a C^∞ function acts as a bounded operator on each Sobolev space. Linear differential operators of order l act boundedly from W^k to W^{k-l}.*

PROOF Obvious. □

COROLLARY 5.11 *Let $f \in L^2(\mathbb{T}^n)$, with support $\mathrm{supp}(f)$ in a compact subset K. Let U be an open subset of $\mathbb{T}^n$ containing K, and let φ be a diffeomorphism of U into $\mathbb{T}^n$. Then $f \circ \varphi$ belongs to W^k if and only if f does.*

PROOF It is enough if we can estimate the L^2 norms of derivatives of $f \circ \varphi$ in terms of the norms of derivatives of f. By the chain rule, the derivatives up to order k of $f \circ \varphi$ can be written as linear combinations of products of derivatives up to order k of f and derivatives up to order k of φ. To compute the L^2 norms of those derivatives one must change variables in the integral, introducing the Jacobian of φ also. However, all the quantities that depend on φ are bounded by compactness, so the result follows. □

Now we can define Sobolev spaces on manifolds. Let M be a compact smooth manifold. Let (U_j) be a cover of M by coordinate patches, and (ψ_j^2) a smooth partition of unity subordinate to U_j. Let φ_j be a diffeomorphism of U_j into $\mathbb{T}^n$.

DEFINITION 5.12 We define the Sobolev k-inner product on $C^\infty(M)$ by

$$\langle f, g \rangle_k = \sum_j \langle (\psi_j f) \circ \varphi_j^{-1}, (\psi_j g) \circ \varphi_j^{-1} \rangle_k .$$

The Sobolev k-inner products on the right hand side refer to $\mathbb{T}^n$. Of course, the norm associated to this inner product depends on the various choices made in its definition. However, (5.10) and (5.11) show that if we make different choices, we replace the norm by an equivalent one. So the k-norm is canonically defined up to equivalence. We define the Sobolev space $W^k(M)$ to be the completion of $C^\infty(M)$ in the k-norm; it is a topological vector space provided with a class of inner products which define equivalent norms, with respect to any of which it is complete (sometimes called a 'Hilbertian space').

If V is a vector bundle over M, one can define similarly the Sobolev space $W^k(V)$ of W^k sections of V, by making an arbitrary choice of trivialization of V over each of the coordinate patches U_j.

Notice that propositions 5.6, 5.7, and 5.8 still apply to the Sobolev spaces of an arbitrary manifold. It is easy to reduce the more general versions of these propositions to their special cases on the torus.

Analysis of the Dirac operator

We now return to the Dirac operator D on a Clifford bundle S over the manifold M. A critical part in the analysis is played by the Weitzenbock formula 3.8, which we recall states that

$$D^2 = \nabla^*\nabla + \mathsf{K}$$

where K is a certain curvature operator. In fact, the precise form of the operator K is of little importance here, and all the analysis will work for any first order operator D on sections of a bundle S (with hermitian metric and compatible connection) which satisfies

$$D^2 = \nabla^*\nabla + B \tag{5.13}$$

where B is a first order operator on S. Such an operator is called a *generalized Dirac operator*; an important example is the operator $D+A$, where D is a Dirac operator in the old sense and A is any endomorphism of S. The Dolbeault operator $\sqrt{2}(\bar{\partial} + \bar{\partial}^*)$ on a non-Kähler complex manifold, for example, is of this form.

Since D is a first order operator, (5.10) gives the estimate

$$\|Ds\|_0 \leqslant C\|s\|_1$$

for some constant C. The main analytical property of generalized Dirac operators is a sort of 'approximate converse' to this:

GARDING'S INEQUALITY: 5.14 Let D be a generalized Dirac operator on a compact manifold. There is a constant C such that, for any $s \in C^\infty(S)$,

$$\|s\|_1 \leqslant C(\|s\|_0 + \|Ds\|_0) \ .$$

PROOF By means of a partition of unity, one can reduce to the case where s is supported in a coordinate patch. Now we use the formula 5.13. Taking the L^2 inner product with s, one gets

$$\|Ds\|_0^2 = \|\nabla s\|_0^2 + \langle Bs, s\rangle_0$$

so, using Cauchy-Schwarz and the fact that B is first order,

$$\|\nabla s\|_0^2 \leqslant C_1(\|s\|_0\|s\|_1 + \|Ds\|_0^2) \tag{5.15}$$

for some constant C_1. Now let us write ∇ in local coordinates as $\nabla_i s = \partial s/\partial x^i + \Gamma_i s$, where s is thought of as a vector-valued function and the Christoffel symbols Γ_i are endomorphisms of S. Then

$$\begin{aligned}\|\nabla s\|_0^2 &= \sum_{i,j}\left\{\int g^{ij}(\frac{\partial s}{\partial x^i}, \frac{\partial s}{\partial x^j}) + 2\int g^{ij}\mathrm{Re}(\frac{\partial s}{\partial x^i}, \Gamma_j s) + \int g^{ij}(\Gamma_i s, \Gamma_j s)\right\}\\ &\geqslant C_2\|s\|_1^2 - C_3\|s\|_0\|s\|_1\end{aligned}$$

and so (using 5.15)

$$\|Ds\|_0^2 \geqslant C_4\|s\|_1^2 - C_5\|s\|_0\|s\|_1$$

for some constants C_2, C_3, C_4, C_5. Now use the fact that given any $\varepsilon > 0$ there is a $K > 0$ such that $ab \leqslant \varepsilon a^2 + Kb^2$ for all $a, b > 0$ to write $C_5\|s\|_0\|s\|_1 \leqslant \frac{1}{2}C_4\|s\|_1^2 + C_6\|s\|_0^2$, and so to deduce

$$\|Ds\|_0^2 \geqslant \tfrac{1}{2}C_4\|s\|_1^2 - C_6\|s\|_0^2 \ .$$

Rearranging this and changing notation slightly, one gets the result. □

There is a generalization of Garding's inequality which relates the Sobolev k-norm of Ds to the Sobolev $(k+1)$-norm of s.

PROPOSITION 5.16 (ELLIPTIC ESTIMATE) *For any $k > 0$ there is a constant C_k such that, for any $s \in C^\infty(S)$,*

$$\|s\|_{k+1} \leqslant C_k(\|s\|_k + \|Ds\|_k) .$$

PROOF The case $k = 0$ is just Garding's inequality. To obtain the more general result, we use induction on k. By a partition of unity, we may assume that s is supported in a coordinate patch. Let ∂_i denote the operator $\partial/\partial x^i$.

From (5.9),

$$\|s\|_{k+1} \leqslant A_1 \sum_i \|\partial_i s\|_k$$

for some constant A_1. Now by induction

$$\|\partial_i s\|_k \leqslant C_{k-1}(\|\partial_i s\|_{k-1} + \|D\partial_i s\|_{k-1}) .$$

But ∂_i is a first order operator, so

$$\|\partial_i s\|_{k-1} \leqslant A_2 \|s\|_k .$$

Also $[D, \partial_i]$ is a first order operator, so

$$\begin{aligned}\|D\partial_i s\|_{k-1} &\leqslant \|\partial_i Ds\|_{k-1} + \|[D, \partial_i]s\|_{k-1} \\ &\leqslant A_2\|Ds\|_k + A_3\|s\|_k .\end{aligned}$$

Therefore

$$\|s\|_{k+1} \leqslant nA_1 C_{k-1}\Big(A_2\|Ds\|_k + (A_2 + A_3)\|s\|_k\Big) ,$$

which yields the result. □

To analyze D, we will think of it as an unbounded operator on the Hilbert space $H = L^2(S)$. Recall that an *unbounded operator* on a Hilbert space H is simply a linear map from a dense subspace of H (called the *domain* of the operator) to H. Such operators need not be continuous; but a basic idea in the theory of unbounded operators is that the closedness of the graph of the operator in $H \oplus H$ acts as a partial substitute for continuity.

DEFINITION 5.17 Let A be an unbounded operator. The *graph* G_A of A is the subspace
$$G_A = \{(x, Ax) : x \in \text{dom}(A)\}$$
of $H \oplus H$.

LEMMA 5.18 *The closure $\overline{G}$ of the graph G of the Dirac operator is also a graph.*

PROOF This is in fact a general property of differential operators, and is based on the existence of a so-called 'formal adjoint' operator $D^\dagger$ such that
$$\langle Ds_1, s_2 \rangle = \langle s_1, D^\dagger s_2 \rangle \tag{5.19}$$
for all smooth sections s_1, s_2 of S. For the classical Dirac operators of Chapter 3, we have proved in 3.11 that $D^\dagger = D$. Suppose now that $\overline{G}$ is not a graph. Then there is a point $(0, y)$ in $\overline{G}$ with $y \neq 0$. That is, there is a sequence (x_j) of smooth sections of S with $x_j \to 0$ and $Dx_j \to y$ in $L^2(S)$. But then, for any smooth s,
$$\langle Dx_j, s \rangle \to \langle y, s \rangle, \qquad \langle x_j, D^\dagger s \rangle \to 0$$
as $j \to \infty$. However $\langle Dx_j, s \rangle = \langle x_j, D^\dagger s \rangle$, so $\langle y, s \rangle = 0$ for all smooth s, and so $y = 0$. □

Since $\overline{G}$ is a graph, it too defines an unbounded operator, denoted $\overline{D}$. The domain of $\overline{D}$ is the collection of all $x \in L^2(S)$ such that there is a sequence (x_j) of smooth sections of S for which $x_j \to x$ in $L^2(S)$ and Dx_j converges in $L^2(S)$. By Garding's inequality (5.14), this domain is precisely the Sobolev space $W^1(S)$.

Suppose that x and y are smooth sections of S, and that $Dx = y$. Then by 5.19, for all smooth sections s,
$$\langle x, D^\dagger s \rangle = \langle y, s \rangle .$$
This equation makes sense for arbitrary $x, y \in L^2(S)$; if it holds one says that the equation $Dx = y$ is satisfied *in the weak sense*. Such a concept can be defined for more general partial differential equations, and for most of them the concept of solvability in the weak sense is a proper generalization of honest solvability; but for the Dirac operator it will turn out that this concept is the same as that of ordinary solvability. To prove this we need some additional concepts.

DEFINITION 5.20 A bounded operator A on $L^2(S)$ is called a *smoothing operator* if there is a smooth *kernel* $k(p,q)$ on $M \times M$, with values $k(p,q) \in \text{Hom}(S_q, S_p)$, such that
$$As(p) = \int_M k(p,q)s(q) \cdot \text{vol}(q) \ .$$
Formally, k is a smooth section of $S \boxtimes S^* := \pi_1^* S \otimes \pi_2^* S^*$, where π_1 and π_2 are the canonical projections of $M \times M$ to M. By differentiation under the integral sign, one sees that the range of a smoothing operator consists of smooth sections.

DEFINITION 5.21 A *Friedrichs' mollifier* for S is a family F_ε, $\varepsilon \in (0,1)$ of self-adjoint smoothing operators on $L^2(S)$ such that
  (i) (F_ε) is a bounded family of operators on $L^2(S)$.
  (ii) $([B, F_\varepsilon])$ extends to a bounded family of operators on $L^2(S)$, for any first order differential operator B on S.
  (iii) $F_\varepsilon \to 1$ in the weak topology of operators on $L^2(S)$. (This means that for all $x, y \in L^2(S)$, $\langle F_\varepsilon x, y \rangle \to \langle x, y \rangle$ as $\varepsilon \to 0$.)

Friedrichs' mollifiers exist (see exercise 5.34). Let us grant that for now, and go on to prove our result on weak solutions of $Dx = y$.

PROPOSITION 5.22 Suppose that $x, y \in L^2(S)$, and that $Dx = y$ weakly. Then $x \in W^1(S) = \text{dom}(\overline{D})$, and $\overline{D}x = y$.

PROOF Let F_ε be a Friedrichs' mollifier, and let $x_\varepsilon = F_\varepsilon x$. Then x_ε is smooth, and we may write for $s \in C^\infty(S)$
$$\begin{aligned} \langle Dx_\varepsilon, s \rangle &= \langle x_\varepsilon, D^\dagger s \rangle \\ &= \langle x, F_\varepsilon D^\dagger s \rangle \\ &= \langle x, D^\dagger F_\varepsilon s \rangle + \langle x, [F_\varepsilon, D^\dagger] s \rangle \\ &= \langle y, F_\varepsilon s \rangle + \langle x, [F_\varepsilon, D^\dagger] s \rangle \ . \end{aligned}$$
So there is a constant C such that
$$|\langle Dx_\varepsilon, s \rangle| \leq C \|s\|$$
uniformly in ε. Since $C^\infty(S)$ is dense in $L^2(S)$, this implies that $\|Dx_\varepsilon\| \leq C$.

Now by Garding's inequality (5.14), $\{x_\varepsilon\}$ forms a bounded subset of the Sobolev space W^1. Therefore, there is a sequence of values $\varepsilon_j \to 0$ such that x_{ε_j} tends to a limit weakly in W^1, by the weak compactness of the unit ball of the Hilbertian space W^1. By Rellich's theorem (5.8), x_{ε_j} tends to its limit in the norm topology of $W^0 = L^2$. By property iii) of Friedrichs' mollifiers, this limit must be x; so $x \in W^1$, as asserted. $\square$

REMARK 5.23 For the benefit of readers familiar with unbounded operator theory, we summarize what we have shown in that language. For simplicity restrict attention to the classical case where $D^\dagger = D$, which means that the operator D is *symmetric*, in the sense of unbounded operator theory. Proposition 5.22 above shows that the domain of the *closure* of D is equal to the domain of the (Hilbert space) *adjoint* of D (they are both equal to W^1), and thus that D is *self-adjoint* in the sense of unbounded operator theory. We now go on to develop a spectral decomposition theory for D; Lemma 5.25 and the subsequent calculation are classical results of unbounded operator theory, due to von Neumann, which we have specialized to the case at hand.

PROPOSITION 5.24 *The kernel of D (i.e., the set of $s \in W^1$ such that $Ds = 0$) consists of smooth sections.*

PROOF Let s belong to the kernel of D; we will prove inductively that $s \in W^k$ for all k, and the result will follow by the Sobolev embedding theorem. Suppose then that it is already known that $s \in W^{k-1}$, and let F_ε be a Friedrichs' mollifier. It is easy to check (from the properties of Friedrichs' mollifiers and the definition of the Sobolev spaces) that F_ε and $[D, F_\varepsilon]$ form bounded families of operators on W^{k-1}. Now by the elliptic estimate

$$\|F_\varepsilon s\|_k \leqslant C_k(\|F_\varepsilon s\|_{k-1} + \|DF_\varepsilon s\|_{k-1}) = C_k(\|F_\varepsilon s\|_{k-1} + \|[D, F_\varepsilon]s\|_{k-1})$$

since $Ds = 0$. Thus $\|F_\varepsilon s\|_k$ is bounded, and since $F_\varepsilon s$ converges in L^2 to s, and suitable subsequence converges weakly in W^k, we deduce that $s \in W^k$ as required. $\square$

Recall that G denotes the graph of D, and $H = L^2(S)$. From now on we will assume that $D = D^\dagger$; all true Dirac operators satisfy this condition.

LEMMA 5.25 *Let $J\colon H \oplus H \to H \oplus H$ denote the map $(x,y) \mapsto (y,-x)$. Then there is an orthogonal direct sum decomposition*

$$H \oplus H = \overline{G} \oplus J\overline{G}\,.$$

PROOF Suppose that $(x,y) \in G^\perp$. This means that for all $s \in C^\infty(S)$,

$$\langle (x,y), (s, Ds)\rangle = 0$$

i.e.

$$\langle x, s\rangle + \langle y, Ds\rangle = 0\,,$$

that is, $Dy + x = 0$ weakly. But then by (5.22), $y \in W^1$, so $(y,-x) \in \overline{G}$, so $(x,y) \in J\overline{G}$. □

(5.26) Now define an operator Q as follows: for any $x \in L^2(S) = H$, let $(Qx, \overline{D}Qx)$ be the orthogonal projection of $(x,0)$ onto $\overline{G}$ in $H \oplus H$. Clearly $Qx \in W^1$, and since $\|x\|^2 = \|Qx\|^2 + \|\overline{D}Qx\|^2$, Garding's inequality shows that Q is bounded as an operator $L^2 \to W^1$. Hence, by Rellich's theorem (5.8), Q is compact when considered as a bounded operator on $L^2(S)$. Clearly it is also self-adjoint, positive, and injective, and has norm $\leqslant 1$.

Now we have the following basic result, which decomposes the operator D into manageable (i.e. finite dimensional) pieces.

THEOREM 5.27 *There is a direct sum decomposition of H into a sum of countably many orthogonal subspaces H_λ. Each H_λ is a finite dimensional space of smooth sections, and is an eigenspace for D with eigenvalue λ. The eigenvalues λ form a discrete subset of $\mathbb{R}$.*

PROOF Consider the compact self-adjoint operator Q defined above. The spectral theorem for such operators (proved in most first courses on functional analysis; see [29] for a comprehensive treatment) says that H can be decomposed into an orthogonal direct sum of finite dimensional eigenspaces for Q, with discrete eigenvalues tending to zero. Since Q is positive and injective, the eigenvalues are in fact strictly positive.

Now let $x \in W^1$ be an eigenvector for Q, with eigenvalue $\rho > 0$. Then by 5.25 there is $y \in W^1$ such that

$$(x, 0) = (Qx, \overline{D}Qx) + (-\overline{D}y, y) = \rho(x, \overline{D}x) + (-\overline{D}y, y)$$

and thus $(\rho - 1)x = \overline{D}y$ and $y = -\rho \overline{D}x$. Putting $\lambda^2 = (1 - \rho)/\rho$ and $z = -(1/\rho\lambda)y$, we have

$$\overline{D}x = \lambda z, \qquad \overline{D}z = \lambda x$$

so $x + z$ and $x - z$ are eigenvectors of $\overline{D}$ with eigenvalues λ and $-\lambda$ respectively. We conclude that Thus H can be written as a direct sum (necessarily orthogonal) of eigenspaces for $\overline{D}$, each eigenspace being a finite dimensional subspace of $W^1(S)$.

To show that the eigenvectors are smooth, notice that an eigenvector for D (with eigenvalue λ) is a member of the kernel of the generalized Dirac operator $D - \lambda$. So 5.24 completes the proof. □

REMARK 5.28 Returning to where this chapter began, observe that if D is the operator $i(d/dx)$ on the circle S^1, the decomposition provided by this theorem is just the Fourier series decomposition of $L^2(S^1)$.

The functional calculus

Let $\sigma(D)$ denote the spectrum (set of eigenvalues) of D. Any section $s \in L^2(S)$ has a 'Fourier expansion' as an orthogonal direct sum

$$\sum_{\lambda \in \sigma(D)} s_\lambda$$

where s_λ is the component of s belonging to the λ-eigenspace of D. It is elementary that $\|s_\lambda\| \leq \|s\|$ for all λ.

PROPOSITION 5.29 *A section $s \in L^2(S)$ is smooth if and only if $\|s_\lambda\| = O(|\lambda|^{-k})$ for each k. (In this case we say that the terms of the expansion are rapidly decreasing.*

PROOF Since s_λ is an eigenvector for D with eigenvalue λ, the elliptic estimate gives the bound $\|s_\lambda\|_k \leqslant C_k \lambda^k \|s\|$ for the Sobolev k-norm. The condition of rapid decay therefore implies that the expansion converges in each Sobolev space. □

If f is a bounded function on $\sigma(D)$, we can define a bounded operator $f(D)$ on $L^2(S)$ by setting $f(D)s = \sum f(\lambda)s_\lambda$ where $s = \sum s_\lambda$ as above; in other words, $f(D)$ is the 'diagonal' operator which acts as multiplication by $f(\lambda)$ on the λ-eigenspace of D. The following proposition is apparent from the discussion above.

PROPOSITION 5.30 *The map $f \mapsto f(D)$ is a unital homomorphism from the ring of the bounded functions on $\sigma(D))$ to $B(H)$. The norm of the operator $f(D)$ is less than or equal to the supremum of $|f|$. If D commutes with an operator A, so does every $f(D)$. Moreover, every $f(D)$ maps $C^\infty(S)$ to $C^\infty(S)$. If $f(x) = xg(x)$, with f and g bounded functions, then $f(D) = Dg(D)$ as bounded operators.*

The argument shows that if f itself is rapidly decreasing, that is $|f(\lambda)| = O(|\lambda|^{-k})$ for each k, then $f(D)$ maps $L^2(S)$ to $C^\infty(S)$. In fact $f(D)$ is actually a *smoothing operator* in this case, that is, given by a smooth kernel. To see this notice that for $\lambda \in \sigma(D)$ the orthogonal projection operator P_λ onto the λ-eigenspace of D is smoothing; indeed, any orthogonal projection whose range is a finite-dimensional space of smooth functions is a smoothing operator. Moreover it is not hard to see (exercise 5.35) that for each k there is an $\ell(k)$ such that the Sobolev k-norm (on $M \times M$) of the smoothing kernel of P_λ is bounded by $C_k \lambda^{\ell(k)}$. Thus, if f is rapidly decreasing, the series

$$f(D) = \sum_\lambda f(\lambda) P_\lambda$$

converges in the topology of smoothing kernels on $M \times M$. To summarize, we have proved

PROPOSITION 5.31 *If f is rapidly decreasing, the associated operator $f(D)$ is a smoothing operator. The map from f to the smoothing kernel of $f(D)$ is continuous, from the space $\mathcal{R}(\mathbb{R})$ of rapidly decreasing functions on $\mathbb{R}$ (equipped with its natural Fréchet topology), to the space of smoothing kernels on $M \times M$.*

REMARK 5.32 It is plain from this discussion that there is an N (depending only on the dimension of M) such that, if $f(\lambda) = O(\lambda^{-N})$, then $f(D)$ has a *continuous* kernel. This will be of some importance later.

Notes

There are many expositions of the theory of linear elliptic operators, a part of which has been presented in this chapter. Our approach owes most to Griffiths and Harris [36]. Instead of making Hilbert space theory central, one can prove the main results by constructing a *parametrix* for D; an operator Q such that $DQ - 1$ and $QD - 1$ are smoothing. This line is followed in [12] and in de Rham's book [24], where parametrices are constructed very explicitly.

One can also construct parametrices by making use of the general theory of pseudo-differential operators, as in [34] and [47]. We have not emphasized pseudo-differential operators in this text, since our main concern is with Dirac operators, which are in some sense more "rigid". However, pseudo-differential operators are invaluable when one needs to discuss deformations of elliptic operators; see for example Atiyah and Singer [9].

We have only defined the Sobolev spaces of positive integer order. Sobolev spaces of negative and fractional order can be defined, as well as Sobolev spaces based on L^p rather than L^2 norms; these are of importance in non-linear problems.

For more on unbounded operators, consult Dunford and Schwartz [29].

Exercises

QUESTION 5.33 Investigate whether elements of the Sobolev space $W^{n/2}(\mathbb{T}^n)$ must be continuous (this is the 'critical case' of the Sobolev embedding theorem). Hint: consider the function $(r, \theta) \mapsto \log(1 - \log r)$ on the unit ball in $\mathbb{R}^2$.

QUESTION 5.34 Show that Friedrichs' mollifiers exist, by following the outline below

(i) Choose a function φ on $\mathbb{R}^n$ which is positive, smooth, compactly supported radially symmetric and has $\int \varphi = 1$; and let $\varphi_\varepsilon(x) = \varepsilon^{-n} \varphi(x/\varepsilon)$. Define F_ε o

$L^2(\mathbb{R}^n)$ by the convolution integral

$$F_\varepsilon s(x) = \varphi_\varepsilon * s(x) = \frac{1}{\varepsilon^n} \int \varphi(\frac{x-y}{\varepsilon}) s(y)\, dy.$$

Prove that the operators F_ε are uniformly bounded on L^2.

(ii) Prove that if s is continuous and compactly supported, then $F_\varepsilon s \to s$ uniformly as $\varepsilon \to 0$.

(iii) Deduce that if $s \in L^2$, then $F_\varepsilon s \to s$ in L^2 as $\varepsilon \to 0$.

(iv) Let $B = a(x)\partial/\partial x^1$. By integration by parts, show that

$$[B, F_\varepsilon]s(x) =$$
$$\frac{1}{\varepsilon^n} \int \varphi(\frac{x-y}{\varepsilon}) \partial_1 a(y) s(y)\, dy + \frac{1}{\varepsilon^{n+1}} \int (a(x) - a(y)) \partial_1 \varphi(\frac{x-y}{\varepsilon}) s(y)\, dy$$

and deduce that the operator norm of $[B, F_\varepsilon]$ is uniformly bounded.

(v) Using the construction above in coordinate patches, and a partition of unity, construct Friedrichs' mollifiers on a compact manifold.

QUESTION 5.35 Let K be a smoothing operator on $L^2(S)$. Prove that the L^2 norm of the smoothing kernel of K is bounded by a multiple of the operator norm of K as an operator from $L^2(S)$ to $C^0(S)$. Hence prove that, if K is the projection P onto the λ-eigenspace of D, the W^k norm of its kernel is bounded by $C_k \lambda^\ell$, for some $\ell > k + n/2$.

QUESTION 5.36 Let D be a Dirac operator. Prove that the operators $F_\varepsilon = \exp(-\varepsilon D^2)$, defined by the functional calculus, form a family of Friedrichs' mollifiers.

QUESTION 5.37 Prove the Fredholm alternative theorem for a Dirac operator D: given a complex number λ, either the equation $Du + \lambda u = 0$ has a non-zero solution or for all v there is a unique solution u to the equation $Du + \lambda u = v$. (Take u and v to be C^∞ sections of S.)

CHAPTER 6

Hodge theory

We have seen that several of the classical Dirac operators are related to complexes (in the sense of homological algebra), such as the de Rham complex. Our analysis of Dirac operators allows us to say something about the cohomology of these complexes.

DEFINITION 6.1 Let M be an n-dimensional compact oriented Riemannian manifold, and let $S_0, S_1, \ldots, S_k$ be a sequence of vector bundles over M, equipped with Hermitian metrics and compatible connections. Suppose given differential operators $d_j \colon C^\infty(S_j) \to C^\infty(S_{j+1})$ in such a way that $d_{j+1} d_j = 0$, i.e. that

$$C^\infty(S_0) \xrightarrow{d} C^\infty(S_1) \xrightarrow{d} C^\infty(S_2) \to \cdots \to C^\infty(S_k)$$

is a complex. It will be called a *Dirac complex* if $S = \bigoplus S_j$ is a Clifford bundle whose Dirac operator D equals $d + d^*$.

By (3.23), the de Rham complex of M is an example of a Dirac complex. De Rham's theorem says that the cohomology of this complex is isomorphic to the usual cohomology (with coefficients $\mathbb{C}$) of the manifold M, as computed in algebraic topology. We will not prove de Rham's theorem here; a proof can be found in [24] or [15].

By (3.27) the Dolbeault complex of a Kähler manifold is also a Dirac complex. The Dolbeault complex of any complex manifold is a 'generalized Dirac complex'; the operator $\bar{\partial} + \bar{\partial}^*$ is a self-adjoint generalized Dirac operator in the sense of the last chapter.

To define the cohomology of a Dirac complex we make no use of the metric. The idea which leads to Hodge Theory is the following one: can we use the metric to choose a canonical representative of each cohomology class? Such a cohomology class is an affine subspace of $C^\infty(S)$, a vector space on which the metric gives a natural L^2 inner product; so it is reasonable to look for the element of smallest norm in a cohomology class. If $\mathcal{C} \subseteq C^\infty(S_j)$ is a cohomology class, then it is an affine subspace whose associated vector subspace is $dC^\infty(S_{j-1})$. Therefore, arguing non-rigorously, we expect a norm-minimizing element α to be perpendicular to $dC^\infty(S_{j-1})$; which translates to say that $d^*\alpha = 0$. Since also $d\alpha = 0$, it must be that $D\alpha = 0$, or equivalently that $D^2\alpha = 0$. In this case one says that α is *harmonic*. So we are led to conjecture that each cohomology class has a harmonic representative.

The problem with this argument (as was already pointed out by Weierstrass in the nineteenth century) is that it assumes, but does not prove, the existence of the desired norm-minimizing element. However, the desired conclusion is in fact true:

THEOREM 6.2 (HODGE THEOREM) *Each cohomology class for a Dirac complex contains a unique harmonic representative. Indeed, the j'th cohomology $H^j(S;d)$ of such a complex is isomorphic as a vector space to the space of harmonic sections of S_j.*

PROOF Let $\mathcal{H}^j$ denote the space of harmonic sections of S_j. Then the $\mathcal{H}$'s form a subcomplex of the Dirac complex with trivial differential:

$$\begin{array}{ccccc} \mathcal{H}^{j-1} & \xrightarrow{0} & \mathcal{H}^j & \xrightarrow{0} & \mathcal{H}^{j+1} \longrightarrow \\ \downarrow \iota & & \downarrow \iota & & \downarrow \iota \\ C^\infty(S_{j-1}) & \xrightarrow{d_{j-1}} & C^\infty(S_j) & \xrightarrow{d_j} & C^\infty(S_{j+1}) \longrightarrow \end{array}$$

We shall prove that the inclusion map ι is a chain equivalence. Define an inverse map $P\colon C^\infty(S_j) \to \mathcal{H}^j$ to be the restriction to $C^\infty(S_j)$ of the orthogonal projection $L^2(S_j) \to \mathcal{H}^j$. Then $P\iota = 1$, and $\iota P = 1 - f(D)$, where

$$f(\lambda) = \begin{cases} 1 & (\lambda \neq 0) \\ 0 & (\lambda = 0) \end{cases}$$

and $f(D)$ is defined by the functional calculus (5.30). Let

$$g(\lambda) = \begin{cases} \lambda^{-2} & (\lambda \neq 0) \\ 0 & (\lambda = 0) \end{cases}.$$

Then g is bounded on the spectrum $\sigma(D)$ of D, so that the *Green's operator* $G = g(D)$ is defined; and $D^2 G = f(D) = 1 - \iota P$. But $D^2 G = (dd^* + d^*d)G = dH + Hd$, where $H = d^*G$; for G commutes with d since D^2 commutes with d. So

$$1 - \iota P = dH + Hd.$$

We deduce that H is a chain homotopy between ιP and 1, so ι is a chain equivalence. □

We can get some immediate consequences from this.

COROLLARY 6.3 *The cohomology of a Dirac complex (over a compact manifold!) is finite-dimensional.*

COROLLARY 6.4 (POINCARÉ DUALITY) *Let M be a compact connected oriented n-manifold. Then the cap product*

$$H^k(M;\mathbb{C}) \otimes H^{n-k}(M;\mathbb{C}) \to H^n(M;\mathbb{C}) \cong \mathbb{C}$$

is a non-degenerate pairing, and so places $H^k(M;\mathbb{C})$ and $H^{n-k}(M;\mathbb{C})$ in duality.

PROOF We use de Rham cohomology; then the cap product on cohomology is induced by the exterior product of differential forms. We must check that if $\mathcal{C} \in H^k$ satisfies $\mathcal{C} \cap \mathcal{C}' = 0$ for all $\mathcal{C}' \in H^{n-k}$, then $\mathcal{C} = 0$. To do this, choose any Riemannian metric, and represent $\mathcal{C}$ by a harmonic form α. Then $*\alpha$ is also a harmonic form, representing a cohomology class $\mathcal{C}'$; so $\mathcal{C} \cap \mathcal{C}'$ is represented by the form $\alpha \wedge *\alpha$. The isomorphism of $H^n(M;\mathbb{C})$ with $\mathbb{C}$ is given by integration; but

$$\int \alpha \wedge *\alpha = 0 \Rightarrow \|\alpha\|^2 = 0 \Rightarrow \alpha = 0.$$

This gives the result. □

The last two corollaries (when applied to the De Rham complex) gave examples of purely topological results proved by analytical methods. Of course there are more geometric means of approaching the same results, by means of Morse Theory for example; but there are examples (6.12) on Kähler manifolds of topological consequences of Hodge theory that seem to be inaccessible by purely topological means.

(6.5) To make full use of Poincaré duality one needs to supplement the analytical understanding of duality in terms of differential forms with a geometrical understanding in terms of homology classes. The simplest examples of homology classes are those defined by *closed submanifolds* of a manifold. Let M be a compact oriented

n-manifold and let C be a closed oriented k-dimensional submanifold. Then C defines a linear functional on cohomology by

$$[C]\colon [\alpha]\in H^k(M) \mapsto \int_C \alpha;$$

it is a simple application of Stokes' theorem to see that $\int_C \alpha$ depends only on the cohomology class of α. We can extend $[C]$ to a functional φ_C on $\Omega^k_{L^2}(M)$ by using the orthogonal projection $P\colon \Omega^k_{L^2} \to \mathcal{H}^k$ (which preserves the cohomology class) and defining

$$\varphi(\alpha) = \int_C P\alpha.$$

Notice that the integral of an L^2 function, or form, over a lower-dimensional submanifold is not well-defined in general; but the smoothing property of P means that $\varphi(\alpha)$ is well defined for each L^2 form α, and moreover that φ is a continuous linear functional on L^2. By Riesz' representation theorem for the dual of a Hilbert space, there is a unique $\beta \in \Omega^k_{L^2}$ such that $\varphi(\alpha) = \langle \alpha, \beta \rangle$ for all α; and, since $P^2 = P = P^*$,

$$\langle \alpha, \beta \rangle = \langle P\alpha, \beta \rangle = \langle \alpha, P\beta \rangle$$

so $P\beta = \beta$, that is, β is harmonic. The dual $*\beta$ represents a cohomology class which has the property that

$$\int_C \alpha = \int_M \alpha \wedge *\beta; \tag{6.6}$$

this cohomology class $[P_C] = [*\beta] \in H^{n-k}(M)$ is called the *Poincaré dual* of C. Although we have used the Riemannian structure to define it, the dual cohomology class is independent of the choice of metric, since it is characterized by the metric-independent equation 6.6.

If C and C' are two closed oriented submanifolds of complementary dimensions (that is, their dimensions sum to n) and in general position, then from 6.6, we have

$$\int_C P_{C'} = \pm \int_{C'} P_C$$

where the $-$ sign appears only if both dimensions are odd. It can be shown that this number is the geometric *intersection number* of the submanifolds C and C', that is, it is the total count of the (necessarily isolated) intersection points of C and C', taken with a sign according to the orientations. (The usual proof of this fact proceeds by relating Poincaré duality to the Thom isomorphism theorem; see [15]

In exercise 6.13 we outline an argument which depends only on the analytical tools we have so far developed.) Because of this geometric interpretation, the bilinear form $(\alpha,\beta) \mapsto \int \alpha \wedge \beta$ on the de Rham cohomology is often called the *intersection form*.

(6.7) Bochner introduced the idea of combining the representation of cohomology by harmonic forms with the Weitzenbock formula (3.8, 3.10) to obtain results on the topological consequences of positive curvature. To carry out his method, one needs to know precisely what is the K term in the Weitzenbock formula. We will work out one example.

LEMMA 6.8 *Let $D = d+d^*$ be the de Rham operator. Then the restriction to 1-forms of the Clifford-contracted curvature operator K appearing in the Weitzenbock formula associated to D is equal to the Ricci curvature operator.*

PROOF Let e_i be an orthonormal frame for TM. Then by definition
$$\mathsf{K} = \tfrac{1}{2}\sum_{i,j} c(e_i)c(e_j)K(e_i,e_j).$$
Here K should in fact denote the curvature operator for the *cotangent* bundle T^*M; but, if we use the metric to identify TM and T^*M, then the compatibility of the connection and metric allows us to identify K with the Riemann curvature operator (see exercise 2.32). Thus
$$\mathsf{K}e_k = \tfrac{1}{2}\sum_{i,j,l} e_i e_j e_l(R(e_i,e_j)e_k,e_l) = \sum_a \mathrm{Ric}_{ka}\, e_a$$
by 3.17. The result follows. □

THEOREM 6.9 (BOCHNER) *Let M be a compact oriented manifold whose first Betti number is nonzero. Then M does not have any metric of positive Ricci curvature.*

PROOF Combine (3.10), (6.2) and (6.8). □

Notes

Hodge theory was introduced by Hodge [41], inspired by the representation of the cohomology of a Riemann surface in terms of holomorphic and anti-holomorphic differentials. Bochner's original paper is Bochner [13]. The method is an extremely important one, particularly in complex geometry; see Griffiths and Harris [36].

Exercises

QUESTION 6.10 Let G be a compact connected Lie group. The *Killing form* of G is the bilinear form B on the Lie algebra $\mathfrak{g}$ defied by

$$B(u_1, u_2) = \mathrm{Tr}(\mathrm{ad}(u_1)\,\mathrm{ad}(u_2))\,,$$

where ad denotes the adjoint representation $\mathfrak{g} \to \mathfrak{gl}(\mathfrak{g})$. G is called *semi-simple* if B is negative definite.

(i) Prove that if G is semi-simple, then the form $-B$ on $\mathfrak{g}$ extends uniquely to a bi-invariant Riemannian metric on G.

(ii) Prove that if ∇ is the Riemannian connection and u is an element of $\mathfrak{g}$ (considered as a left invariant vector field on G) then $\nabla_u u = 0$. Deduce that $\nabla_u v = \frac{1}{2}[u,v]$ for $u, v \in \mathfrak{g}$.

(iii) Prove that the Riemann curvature is given by

$$R(u,v)w = -\frac{1}{4}[[u,v],w]$$

for $u, v, w \in \mathfrak{g}$.

(iv) Prove that G has positive Ricci curvature, and deduce that the first Betti number of G is zero. (It may be helpful to know that one-parameter subgroups of a Lie group are geodesics of any bi-invariant metric.)

QUESTION 6.11 Improve Bochner's theorem by showing that the first cohomology vanishes if the Ricci curvature is everywhere non-negative, and positive at just one point. Must $H^1(M)$ vanish if M has zero Ricci curvature everywhere?

QUESTION 6.12 This question gives a simple topological obstruction to the existence of a Kähler metric.

(i) Let R be a compact Riemann surface, equipped with a Riemannian metric that is compatible with its conformal structure. Show that every harmonic 1-form on R is the sum of a holomorphic and an anti-holomorphic 1-form.

(ii) More generally, let M be a compact Kähler manifold. There are then three "Laplacians" that can be defined on the space of differential forms on M

$\Delta_D = (d+d^*)^2$, $\Delta_\partial = (\partial + \partial^*)^2$, and $\Delta_{\bar\partial} = (\bar\partial + \bar\partial^*)^2$. Prove that $\Delta_{\bar\partial} = \Delta_\partial = \frac{1}{2}\Delta_D$ (see [**36**, page 115]).

(iii) Deduce that the conjugate of a harmonic form of type (p,q) is a harmonic form of type (q,p), and hence that the odd Betti numbers of a Kähler manifold are even.

QUESTION 6.13 Let C be a closed oriented k-dimensional submanifold of M as in (6.5). Let g be a rapidly decreasing function on $\mathbb{R}^+$, with $g(0) = 1$, and define a k-form β_g on M by
$$\langle \alpha, \beta_g \rangle = \int_C g(\Delta)\alpha.$$

(i) Prove that β_g is a smooth k-form.

(ii) Prove that $*\beta_g$ is closed, and that its cohomology class does not depend on the choice of g. Deduce that for all g, $*\beta_g$ represents the Poincaré dual of C.

(iii) Prove that given any neighbourhood U of C in M, one can choose g in such a way that $*\beta_g$ is supported within U. (You will probably need to use the results on finite propagation speed from the next chapter.)

(iv) Let U be a *tubular neighbourhood* of C, diffeomorphic to the total space of the normal bundle of C in M. Prove that the integral of $*\beta_g$ over any fiber of the normal bundle is 1 (if orientations are chosen consistently). [Use a local product metric.]

(v) Deduce that if C and C' are submanifolds of complementary dimension meeting transversally, then the integral $\int P_C \wedge P_{C'}$ is equal to their geometric intersection number.

CHAPTER 7

The heat and wave equations

In this chapter we will study two important partial differential equations involving the Dirac operator. So, throughout the chapter, M will denote a compact Riemannian manifold, equipped with a Dirac operator D acting on sections of a Clifford bundle S. (The reader is invited to check that all the results remain valid for the larger class of generalized Dirac operators of the form $D + A$, where A is a self-adjoint endomorphism of S.)

Existence and uniqueness theorems

DEFINITION 7.1 The *heat equation* for D is the partial differential equation

$$\frac{\partial s}{\partial t} + D^2 s = 0 \qquad (7.2)$$

The *wave equation* for D is the partial differential equation

$$\frac{\partial s}{\partial t} - iDs = 0 \qquad (7.3)$$

In both cases, s is a smooth section of S depending smoothly on the "time" parameter t; we shall often write s as $t \mapsto s_t$, where s_t is a smooth section of S.

PROPOSITION 7.4 *Both the heat and wave equations have unique smooth solutions corresponding to given smooth initial data s_0. The solutions exist for all $t \geqslant 0$ in the case of the heat equation, and for all $t \in \mathbb{R}$ in the case of the wave equation. They satisfy L^2 norm estimates of the form $\|s_t\| \leqslant \|s_0\|$ in the case of the heat equation, $\|s_t\| = \|s_0\|$ in the case of the wave equation.*

PROOF We do it for the heat equation. First of all, assume that there is a smooth

solution s_t. Then
$$\begin{aligned}\frac{\partial}{\partial t}\|s_t\|^2 &= \frac{\partial}{\partial t}\langle s_t, s_t\rangle \\ &= -\langle D^2 s_t, s_t\rangle - \langle s_t, D^2 s_t\rangle \\ &= -2\|Ds_t\|^2 \leq 0.\end{aligned}$$

This gives the *a priori* estimate
$$\|s_t\|^2 \leq \|s_0\|^2 \qquad (t \geq 0)$$
and proves uniqueness. Now for existence, put
$$s_t = e^{-tD^2} s_0 \tag{7.5}$$
where the operator e^{-tD^2} is defined by functional calculus (5.30); then $s_t \in C^\infty(S)$. Moreover, since the function $\lambda \mapsto e^{-t\lambda^2}$ can be differentiated arbitrarily often with respect to $t > 0$, uniformly in λ, we may differentiate (7.5) to find that s_t depends smoothly on t and $\partial s_t/\partial t = -D^2 s_t$. So s_t is a solution.

The proof for the wave equation is similar. Notice that the uniqueness proof applies to any solution which is C^2 in space and C^1 in time. □

From the functional calculus, the solution operator e^{-tD^2} to the heat equation is a smoothing operator. Thus there is a time-dependent section k_t of the bundle $S \boxtimes S^*$ over $M \times M$, called the *heat kernel*, such that
$$e^{-tD^2} s(p) = \int_M k_t(p, q) s(q) \, \text{vol}(q)$$
for all smooth sections s and all $t > 0$.

PROPOSITION 7.6 *The heat kernel $k_t(p,q)$ has the following properties.*

(i) *We have*
$$\left[\frac{\partial}{\partial t} + D_p^2\right] k_t(p,q) = 0$$
where D_p denotes the Dirac operator D applied in the p-variable. That is, for each fixed q, the section $p \mapsto k_t(p,q)$ of $S \otimes S_q^*$ satisfies the heat equation.

(ii) *For each smooth section s,*
$$\int_M k_t(p,q) s(q) \, \text{vol}(q) \to s(p)$$

uniformly in p as $t \to 0$.

Moreover, the heat kernel is the unique time-dependent section of $S \boxtimes S^*$ which is C^2 in p and q, C^1 in t, and has the properties (i) and (ii) above.

PROOF It is plain from our discussion above that the heat kernel has properties (i) and (ii).

Conversely, suppose that k_t has these properties, and let K_t be the family of smoothing operators with kernels k_t. Then, for any section s, the time-dependent section $K_t s$ satisfies the heat equation for $t > 0$; and therefore, by uniqueness for C^2 solutions of the heat equation,

$$K_t s = e^{-(t-\varepsilon)D^2} K_\varepsilon s$$

for all $\varepsilon > 0$. But, as $\varepsilon \to 0$, $K_\varepsilon s \to s$ uniformly, and $e^{-(t-\varepsilon)D^2} \to e^{-tD^2}$ in L^2 operator norm; so $e^{-(t-\varepsilon)D^2} K_\varepsilon s \to e^{-tD^2} s$. It follows that $K_t s = e^{-tD^2} s$ for all s, so K_t is the heat kernel. □

REMARK 7.7 Property (ii) of the heat kernel may be expressed by saying that k_t 'tends to a δ-function' as $t \to 0$.

Our proof of the Index Theorem will be based on the study of certain approximations to the heat kernel.

DEFINITION 7.8 Let m be a positive integer. An *approximate heat kernel* of order m is a time-dependent section $k'_t(p,q)$ of $S \boxtimes S^*$ which is C^1 in t, C^2 in p and q, and which tends to a δ-function in the sense of property (ii) above, and which approximately satisfies the heat equation in the sense that

$$\left[\frac{\partial}{\partial t} + D_p^2 \right] k'_t(p,q) = t^m r_t(p,q)$$

where $r_t(p,q)$ is a C^m section of $S \boxtimes S^*$ and depends continuously on t for $t \geq 0$.

We aim to prove that approximate heat kernels are asymptotic, in an appropriate sense, to the true heat kernel. Our main tool is the following result, known as Duhamel's principle.

97

PROPOSITION 7.9 Let s_t be a continuously varying C^2 section of S. Then there is a unique smooth section $\tilde{s}_t$ of S, differentiable in t and with $\tilde{s}_0 = 0$, satisfying the inhomogeneous heat equation

$$\left[\frac{\partial}{\partial t} + D^2\right]\tilde{s}_t = s_t.$$

In fact, $\tilde{s}_t$ is given by the integral formula:

$$\tilde{s}_t = \int_0^t e^{-(t-t')D^2} s_{t'}\, dt'.$$

PROOF Uniqueness follows from uniqueness in the ordinary heat equation (7.4). As for existence, differentiate the formula, getting

$$\begin{aligned}\frac{\partial \tilde{s}_t}{\partial t} &= s_t + \int_0^t (-D^2 e^{-(t-t')D^2} s_{t'})\, dt' \\ &= s_t - D^2 \tilde{s}_{t'} \quad \square\end{aligned}$$

COROLLARY 7.10 For each $k \geq 0$ there are estimates in Sobolev norms for the solution of the inhomogeneous heat equation 7.9 of the form

$$\|\tilde{s}_t\|_k \leq tC_k \sup\{\|s_{t'}\|_k : 0 \leq t' \leq t\}.$$

PROOF This follows from the integral formula once we know that the operators e^{-tD^2} are uniformly bounded on every Sobolev space. This fact is a consequence, by an argument which is surely familiar by now, of the elliptic estimate (5.16), the uniform boundedness of e^{-tD^2} on L^2 (5.30), and the fact that D^k commutes with e^{-tD^2}. $\square$

PROPOSITION 7.11 Let k_t denote the true heat kernel on M. For every m there exists an $m' \geq m$ such that, if k_t' is an approximate heat kernel of order m', then

$$k_t(p,q) - k_t'(p,q) = t^m e_t(p,q)$$

where e_t is a C^m section of $S \boxtimes S^*$ depending continuously on $t \geq 0$.

PROOF We take $m' > m + \frac{1}{2} \dim M$. By definition, the approximate heat kernel k'_t tends to a δ-function and satisfies $(\partial/\partial t + D_p^2)k'_t(p,q) = t^m r_m(p,q)$, where r_m is a C^m error term. Let $s_t(p,q)$ denote the unique solution (dependent on q) to the inhomogeneous heat equation

$$(\frac{\partial}{\partial t} + D_p^2)s_t(p,q) = -t^m r_t(p,q)$$

with $s_0 = 0$. Then the uniqueness of the heat kernel shows that

$$k'_t(p,q) + s_t(p,q) = k_t(p,q).$$

But by (7.10), $\|s_t\|_{m'} \leqslant Ct^{m'+1}$ for some constant C. The Sobolev embedding theorem now completes the proof. $\square$

The asymptotic expansion for the heat kernel

Now we will show how to build an approximate heat kernel from local data. Recall from the elementary theory of PDE that the heat kernel on Euclidean space is the function

$$(x,y,t) \mapsto \frac{1}{(4\pi t)^{n/2}} \exp\{-|x-y|^2/4t\}.$$

This suggests consideration of the function

$$h_t(p,q) = \frac{1}{(4\pi t)^{n/2}} \exp\{-d(p,q)^2/4t\}.$$

on a Riemannian manifold M as a first approximation to the heat kernel there.

Let us fix the point q, and take a geodesic local coordinate system x^i with q as origin. Let $r^2 = \sum (x^i)^2 = \sum g^{ij} x^i x^j$, so that r is the geodesic distance from q and h is the function $(4\pi t)^{-n/2} e^{-r^2/4t}$.

LEMMA 7.12 *We have the following expressions for the derivatives of the function h:*
(a) $\nabla h = -\dfrac{h}{2t} r \dfrac{\partial}{\partial r}$;

(b) $\dfrac{\partial h}{\partial t} + \Delta h = \dfrac{rh}{4gt} \dfrac{\partial g}{\partial r}$, *where $g = \det(g_{ij})$ is the determinant of the metric.*

PROOF It is clear that $dh = (-h/2t) r dr$. The gradient ∇h is the vector field that corresponds to dh under the isomorphism provided by the metric between the tangent and cotangent spaces; but this isomorphism takes dr to $\partial/\partial r$ because the coordinate system is geodesic. This proves (a).

As for (b), we recall the general formula
$$\nabla^*(fV) = f\nabla^*V - \langle \nabla f, V\rangle$$
for a function f and vector field V. Now $\Delta h = \nabla^*\nabla h$, and we have already computed ∇h in (a). Thus we get
$$\Delta h = -\frac{h}{2t}\nabla^*\left(r\frac{\partial}{\partial r}\right) + \frac{r}{2t}\frac{\partial h}{\partial r}.$$
The second term equals $-r^2 h/4t^2$. To compute the first we may use the formula 1.26 to write
$$\nabla^*\left(r\frac{\partial}{\partial r}\right) = -\frac{1}{\sqrt{g}}\sum_j \frac{\partial}{\partial x^j}\left(x^j\sqrt{g}\right) = -n - \frac{r}{2g}\frac{\partial g}{\partial r}.$$
Thus
$$\Delta h = \left(-\frac{r^2}{4t^2} + \frac{n}{2t} + \frac{r}{4gt}\frac{\partial g}{\partial r}\right)h.$$
On the other hand, it is easy to work out that $\partial h/\partial t = (-n/2t + r^2/4t^2)h$, and combining this with our calculation of Δh we get the result. $\square$

We will also need some calculations about the commutator of the Dirac operator with multiplication by a smooth function.

LEMMA 7.13 *Let D be the Dirac operator on sections of a Clifford bundle S, let s be a section of S, and let f be a smooth function. Then*

(a) $D(fs) - fDs = c(\nabla f)s$, *where c denotes Clifford multiplication.*
(b) $D^2(fs) - fD^2s = (\Delta f)s - 2\nabla_{\nabla f}s$.

PROOF Choose a synchronous orthonormal frame e_i. Then we have
$$D(fs) = \sum_i e_i\nabla_i(fs) = f\sum_i e_i\nabla_i s + \sum_i df(e_i)e_i \cdot s = fDs + c(\nabla f)s.$$
This proves (a). For (b), a similar computation gives
$$D^2(fs) = f\sum_{i,j} e_i e_j \nabla_i \nabla_j s + \sum_{i,j}(\nabla_i\nabla_j f)e_i e_j s + \sum_{i,j} e_i e_j \left[\nabla_i f \nabla_j s + \nabla_j f \nabla_i s\right].$$
The first term is equal to $fD^2 s$, the second to $(\Delta f)s$, and in the third the terms with $i \neq j$ cancel to leave simply $-2\sum \nabla_i f \nabla_i s = -2\nabla_{\nabla f}s$. $\square$

Our intention now is to use the function h_t as a starting-point for the construction of an *asymptotic expansion* for the heat kernel.

DEFINITION 7.14 Let f be a function on $\mathbb{R}^+$ with values in a Banach space E. A formal series

$$f(t) \sim \sum_{k=0}^{\infty} a_k(t),$$

where the a_k are functions $\mathbb{R}^+ \to E$, is called an *asymptotic expansion* for f near $t = 0$ if for each positive integer n there exists an ℓ_n such that, for all $\ell \geqslant \ell_n$ there is a constant $C_{\ell,n}$ such that

$$\left\| f(t) - \sum_{k=0}^{\ell} a_k(t) \right\| \leqslant C_{\ell,n} |t|^n$$

for sufficiently small t.

To put this more straightforwardly, for any n, almost all the partial sums of the series must approximate f to within an error which is of order t^n. An asymptotic expansion need not converge: an instructive example is furnished by the Maclaurin series

$$f(t) \sim \sum_{k=0}^{\infty} \frac{f^{(k)}(0)}{k!} t^k$$

for a C^∞ function. This is always a valid asymptotic expansion; but it is convergent only if f is analytic near zero.

We will obtain the following asymptotic expansion for the heat kernel.

THEOREM 7.15 *Let M be a compact Riemannian manifold equipped with a Clifford bundle S and Dirac operator D. Let k_t denote the heat kernel of M. Then*

(i) *There is an asymptotic expansion for k_t, of the form*

$$k_t(p, q) \sim h_t(p, q)(\Theta_0(p, q) + t\Theta_1(p, q) + t^2\Theta_2(p, q) + \cdots),$$

where the Θ_j are smooth sections of $S \boxtimes S^*$.

(ii) *The expansion is valid in the Banach space $C^r(S \boxtimes S^*)$ for all[1] integers $r \geqslant 0$. It may be differentiated formally to obtain asymptotic expansions for the derivatives (both with respect to x and t) of the heat kernel.*

---

[1] The constants implicit in the notion 'asymptotic expansion' may depend on r, of course.

(iii) *The values* $\Theta_j(p,p)$ *of the sections* Θ_j *along the diagonal can be computed by algebraic expressions involving the metrics and connection coefficients, and their derivatives, of which the first is* $\Theta_0(p,p) = 1$, *the identity endomorphism of* S.

PROOF Because of proposition 7.11, it suffices to show that one can determine smooth sections Θ_j of $S \boxtimes S^*$ in such a way that for each m the partial sum

$$h_t(p,q) \sum_{j=0}^{J} t^j \Theta_j(p,q)$$

is an approximate heat kernel of order m for all sufficiently large J. Moreover, it suffices to construct the $\Theta_j(p,q)$ for p near to q, since $h_t(p,q)$ is of order t^∞ outside any neighbourhood of the diagonal in $M \times M$. We may therefore use local coordinates, so fix a geodesic local coordinate system with origin q, and let $x^1, \ldots, x^n$ be local coordinates for the point p. Let h be the local coordinate representation of the smooth function $h_t(\cdot, q)$ defined above.

By 7.12 and 7.13, we have for any section s of S (or of $S \otimes S_q^*$)

$$\frac{1}{h}\left[\frac{\partial}{\partial t} + D^2\right](hs) = \frac{\partial s}{\partial t} + D^2 s + \frac{r}{4gt}\frac{\partial g}{\partial r}s + \frac{1}{t}\nabla_{r\partial/\partial r}s. \tag{7.16}$$

Now write $s \sim u_0 + tu_1 + t^2 u_2 + \ldots$ where the u_j are independent of t, and attempt to solve the equation $(\partial/\partial t - L)(hs) = 0$ by equating to zero the coefficients of powers of t in 7.16 above. We obtain the following system of equations for $j = 0, 1, 2, \ldots$:

$$\nabla_{r\partial/\partial r} u_j + \left(j + \frac{r}{4g}\frac{\partial g}{\partial r}\right)u_j = -D^2 u_{j-1}. \tag{7.17}$$

The equations 7.17 are just ordinary differential equations along each ray emanating from the origin, and we may solve them recursively. To do this we introduce an integrating factor $g^{1/4}$ and rewrite the equations as

$$\nabla_{\partial/\partial r}\left(r^j g^{1/4} u_j\right) = \begin{cases} 0 & (j=0), \\ -r^{j-1} g^{1/4} D^2 u_{j-1} & (j \geq 1) \end{cases}$$

For $j = 0$, this shows that u_j is uniquely determined by its initial value $u_j(0)$, which we fix as 1, the identity endomorphism of S_q. For $j \geq 1$ the equation determines u in terms of u_{j-1}, up to the addition of a constant multiple of a term which is of order

r^{-j} near $r = 0$. The requirement of smoothness at the origin forces this constant of integration to vanish, so we conclude that all the u_j are uniquely determined by the single initial condition $u_0(0) = 1$.

Now define $\Theta_j(p,q)$ to be the $S \boxtimes S^*$-valued function which is represented in local coordinates near q by the function $u_j(x)$ constructed in the previous paragraph. Since $\Theta_0(p,p) = 1$, elementary estimates show that for any J the partial sum

$$k_t^J(p,q) = h_t(p,q) \sum_{j=0}^{J} t^j \Theta_j(p,q)$$

tends to a δ-function as $t \to 0$. Moreover the construction of the u_j shows that

$$\left[\frac{\partial}{\partial t} + D_p^2\right] k_t^J(p,q) = t^J h_t(p,q) e_t^J(p,q)$$

for some smooth error term $e_t^J(p,q)$. But for $J > m + n/2$, the function $t^J h_t(p,q)$ tends to zero in the C^m topology as $t \to 0$. Thus, for sufficiently large J, $k_t^J(p,q)$ is an approximate heat kernel of order m. As we already observed, this together with 7.11 establishes that

$$h_t(p,q) \sum_j t^j \Theta_j(p,q)$$

is an asymptotic expansion for the heat kernel, as required.

Finally, we must justify our assertion that the $\Theta_j(p,p)$ can be computed by algebraic expressions involving the metric and connection coefficients and their derivatives. In terms of our local expressions, $\Theta_j(p,p)$ corresponds to $u_j(0)$; and notice that the coefficients in the differential equation 7.17 are themselves functions of the metric and connection coefficients of the sort described. Expand both sides of 7.17 in Taylor series about the origin, and compare coefficients. The assertion follows by induction on j. □

EXAMPLE 7.18 In principle, it is possible to calculate all the coefficients Θ_j in the asymptotic expansion just by following through the proof. But in practice the details soon become exhausting. The computation of the second term is not too laborious, however; we will compute the coefficient Θ_1 along the diagonal.

From the calculations above, $u_0 = g^{-1/4}$. We substitute this back into 7.17 to find u_1. We only want the value of u_1 at the origin, and from 7.17 this is given by

$$u_1(0) = -D^2 u_0(0) = \sum_i \left(\frac{\partial}{\partial x^i}\right)^2 (g^{-1/4}) - \mathsf{K} \quad \text{at the origin,}$$

where K is the Clifford-contracted curvature term which appears in the Weitzenbock formula. Now from the Taylor expansion of the metric in geodesic coordinates (1.32), $g = 1 + \frac{1}{3}\sum x^p x^q R_{ipqi} + O(|x|^3)$, so $g^{-1/4} = 1 - \frac{1}{12}\sum x^p x^q R_{ipqi} + O(|x|^3)$. Therefore

$$\sum_i \left(\frac{\partial}{\partial x^i}\right)^2 (g^{-1/4}) = -\sum_{i,p} \frac{1}{6} R_{ippi} = \frac{1}{6}\kappa$$

where κ denotes the scalar curvature (1.13).

We state our result formally as a proposition:

PROPOSITION 7.19 *The asymptotic expansion for the heat kernel of D^2 begins with the terms*

$$\begin{aligned}\Theta_0(p,p) &= 1 \\ \Theta_1(p,p) &= \tfrac{1}{6}\kappa(p) - \mathsf{K}(p),\end{aligned}$$

where $\kappa(p)$ denotes the scalar curvature at the point p and $\mathsf{K}(p)$ is the Clifford-contracted curvature operator appearing in the Weitzenbock formula.

Finite propagation speed for the wave equation

A consequence of the asymptotic expansion is that as $t \to 0$ the heat kernel becomes more and more localized near the diagonal in $M \times M$. In this section we'll discuss another method, using the wave equation, of obtaining such localization results. The method is particularly useful in the study of *non*-compact manifolds, as we will see later. It is based on a fundamental fact about the wave equation: disturbances governed by it ("photons") travel at a finite speed. In fact, with the normalization we have selected, this speed is 1.

PROPOSITION 7.20 *For any $s \in C_c^\infty(S)$, the support of $e^{itD}s$ lies within a distance $|t|$ of the support of s.*

PROOF This will be done by means of an "energy estimate", as it is called. The energy estimate is the following

CLAIM: Let $m \in M$, and let $B(m; r)$ denote the ball of radius r around m. Choose R sufficiently small that $B(m; R)$ is contained in the domain of a geodesic co-ordinate system around m. Let s_t be a solution to the wave equation. Then the integral

$$\int_{B(m;R-t)} |s_t|^2$$

is a decreasing function of t.

Before proving this claim, we check that it implies the stated result. To prove the result in general it is enough to prove it for t small and positive, because of the group property $e^{it_1 D} e^{it_2 D} = e^{i(t_1+t_2)D}$ and the duality $(e^{itD})^* = e^{-itD}$. Choose R small enough that for any $m \in M$, the ball $B(m; R)$ lies in the domain of a geodesic co-ordinate system around m. Then for all m at a distance R or more from supp(s),

$$\int_{B(m;R)} |s|^2 = 0.$$

Therefore, for $0 < t < R$,

$$\int_{B(m;R-t)} |e^{itD} s|^2 = 0$$

by the claim, and hence in particular $e^{itD} s(m) = 0$. This proves the result.

To check the claim, differentiate the expression $\int_{B(m;R-t)} |s_t|^2$ with respect to t, obtaining

$$\int_{B(m;R-t)} [(iDs_t, s_t) + (s_t, iDs_t)] - \int_{S(m;R-t)} (s_t, s_t) \, d\sigma \qquad (7.21)$$

where S denotes the sphere and $d\sigma$ is the element of surface area on S. Now recall from the proof of (3.11) that

$$(iDs_t, s_t) + (s_t, iDs_t) = id^*\omega$$

where ω is the 1-form $\omega(X) = -(Xs_t, s_t)$. Therefore, by the divergence theorem, the first term of (7.21) is equal to

$$-i \int_{S(m;R-t)} (N.s_t, s_t) \, d\sigma$$

where N is the unit normal to S. By Cauchy-Schwarz,

$$\left| \int_{S(m;R-t)} (N.s_t, s_t) d\sigma \right| \leq \int_{S(m;R-t)} |s_t|^2 \, d\sigma,$$

105

since Clifford multiplication by the unit vector N acts as an isometry. The claim follows. □

We will need some properties of the Fourier transform on Schwartz space. Recall that the *Schwartz space* $S(\mathbb{R})$ is the space of C^∞ functions on $\mathbb{R}$ which are rapidly decreasing and all of whose derivatives are also rapidly decreasing.

If $f \in S(\mathbb{R})$ its *Fourier transform* $\hat{f}$ is defined by

$$\hat{f}(\lambda) = \int f(x) e^{-ix\lambda}\, dx.$$

The *Fourier inversion formula*

$$f(x) = \frac{1}{2\pi} \int \hat{f}(\lambda) e^{+ix\lambda}\, d\lambda$$

shows that the Fourier transformation gives a linear homeomorphism of $S(\mathbb{R})$ to $S(\hat{\mathbb{R}})$. The theory of the Fourier transformation may be found e.g. in Rudin [64].

Now let $f \in S(\mathbb{R})$ and consider the operator $f(D)$ defined by the functional calculus 5.30. We may write

$$f(D) = \frac{1}{2\pi} \int \hat{f}(\lambda) e^{i\lambda D}\, d\lambda. \tag{7.22}$$

The vector-valued integral should be thought of in the 'weak sense' that

$$\langle f(D)x, y \rangle = \frac{1}{2\pi} \int \hat{f}(\lambda) \langle e^{i\lambda D} x, y \rangle\, d\lambda$$

for all $x, y \in L^2(S)$. To prove this it is enough (by the spectral theorem) to consider the case when x, y are eigenvectors of D; but there it reduces to the Fourier inversion formula.

Now combine the formula 7.22 with the unit propagation speed of the wave equation. We obtain

PROPOSITION 7.23 *Suppose that $f \in S(\mathbb{R})$ and that the Fourier transform $\hat{f}$ is supported in $[-c, c]$. Then $\langle f(D)x, y \rangle = 0$ whenever x and y are sections of S whose supports satisfy* $\mathrm{d}(\mathrm{supp}(x), \mathrm{supp}(y)) > c$. *Consequently, the smoothing kernel of $f(D)$ is supported within a c-neighbourhood of the diagonal in $M \times M$.*

We can easily derive a localization property, which includes that for the heat kernel as a special case.

PROPOSITION 7.24 Let $f \in \mathcal{S}(\mathbb{R})$. As $u \to 0$, the smoothing kernel of the operator $f(uD)$ tends to zero on the complement of any neighbourhood of the diagonal in $M \times M$.

PROOF Given a neighbourhood of the diagonal, choose $\delta > 0$ such that all points within distance 2δ of the diagonal lie within the given neighbourhood. Then pick a C^∞ "bump function" ψ on $\mathbb{R}$ such that

$$\psi(\lambda) = \begin{cases} 1 & (|\lambda| < \delta) \\ 0 & (|\lambda| > 2\delta) \end{cases}$$

and let f_1, f_2 be the Schwartz-class functions whose Fourier transforms are

$$\hat{f}_1(\lambda) = (1/u)\hat{f}(\lambda/u)\psi(\lambda), \quad \hat{f}_2(\lambda) = (1/u)\hat{f}(\lambda/u)(1 - \psi(\lambda)).$$

Then $f_1(x) + f_2(x) = f(ux)$. The operators $f_1(D)$ and $f_2(D)$ are smoothing, and by proposition 7.23, the kernel of $f_1(D)$ is supported within 2δ of the diagonal. Outside the given neighbourhood of the diagonal, therefore, the kernel of $f(uD)$ is equal to the kernel of $f_2(D)$. But as $t \to 0$, $\hat{f}_2 \to 0$ in the Schwartz space $\mathcal{S}(\mathbb{R})$, so by Fourier theory $f_2 \to 0$ in $\mathcal{S}(\mathbb{R})$, and so the smoothing kernel of $f_2(D)$ tends to zero in the C^∞ topology. □

Notes

The heat and wave equations are, of course, standard topics in Mathematical Physics. Our "wave equation" is in a sense the square root of the usual one; Dirac operators were introduced in order that such a square root could be extracted.

The main result of this chapter — the asymptotic expansion of the heat kernel — has a long history. It was first proved (for the scalar Laplacian) by Minakshisundaram and Pleijel [56], and generalized to the Laplacian on differential forms and other operators by McKean and Singer [51], Gilkey [33] and Patodi [58]. We have followed more or less the argument of Patodi's paper. A direct proof can also be given using the calculus of pseudo-differential operators; see [34].

Our proof of the finite propagation speed for the wave equation is modeled on that of Chernoff [19]. The observation that finite propagation speed together with the

Fourier transform can be used to obtain kernel estimates is due to Cheeger, Gromov and Taylor [18].

Exercises

QUESTION 7.25 Use the Poisson summation formula to show that the heat kernel on a flat torus agrees with that on Euclidean space up to an exponentially small error.

QUESTION 7.26 A function f on $\mathbb{R}$ is said to belong to the class $S^0(\mathbb{R})$ if it is smooth and satisfies estimates of the form

$$|f^{(k)}(x)| \leqslant C_k(1+|x|)^{-k}.$$

Prove that if f belongs to this class and D is a Dirac operator, then $f(D)$ has the following pseudo-local property: for any s belonging to $L^2(S)$, the section $f(D)s$ is smooth on the complement of the support of s. (Use the Fourier representation and (7.20)). In fact $f(D)$ is an example of a pseudo-differential operator; see Taylor [71, Chapter XII].

CHAPTER 8

Traces and eigenvalue asymptotics

Eigenvalue growth

Let M be a compact oriented n-dimensional manifold. We have seen in chapter 5 that the Laplacian operator Δ on $L^2(M)$ has discrete spectrum, with eigenvalues $0 \leqslant \lambda_0 \leqslant \lambda_1 \leqslant \lambda_2 \leqslant \cdots$ tending to infinity. In this chapter we want to refine this result by asking *how many* eigenvalues of Δ lie below a fixed value of λ; in other words, we want to study the *counting function*

$$\mathfrak{N}(\lambda) = \max\{j : \lambda_j \leqslant \lambda\}.$$

The results we obtain will also be valid (with trivial modifications) for the square D^2 of any generalized Dirac operator.

A crude estimate for the counting function can be obtained from the Sobolev embedding theorem. Specifically, let $s_1, \ldots, s_j$, $j = \mathfrak{N}(\lambda)$, be orthornormalized eigenfunctions belonging to eigenvalues $\leqslant \lambda$. Let $s = \sum \alpha_i s_i$ be any linear combination of $s_1, \ldots, s_j$. Using the elliptic estimates and the Sobolev embedding theorem we find that there is a constant C (depending only on the geometry) such that for all $x \in M$

$$|s(x)| \leqslant C(1+\lambda)^{k/2} \left(\sum |\alpha_i|^2\right)^{1/2}$$

where k is the least integer strictly greater than $n/2$. Fix $x \in M$, take $\alpha_i = \bar{s}_i(x)$, and rearrange to obtain the identity

$$\sum |s_i(x)|^2 \leqslant C^2(1+\lambda)^k.$$

Integrate over M to obtain $j = \mathfrak{N}(\lambda) \leqslant C^2(1+\lambda)^k \operatorname{vol}(M)$.

A more precise estimate of $\mathfrak{N}(\lambda)$ can be obtained from the heat equation asymptotics of the previous chapter. The link between dimensions (of eigenspaces) and functional analysis is provided by the notion of *trace* for appropriate compact operators on a Hilbert space. Simply put, the trace of an operator is the sum of the diagonal entries of an infinite matrix representing it; and, as in ordinary linear algebra, the trace of a *projection* is the dimension of its range. We begin with the general theory of traces.

Trace-class operators

Let H and H' be (separable, infinite dimensional) Hilbert spaces, and choose orthonormal bases (e_i) and (e'_j) in H and H'. A bounded linear operator $A\colon H \to H'$ can be represented by an "infinite matrix" with coefficients[1]

$$c_{ij}(A) = \langle Ae_i, e'_j \rangle.$$

PROPOSITION 8.1 *The quantity*

$$\|A\|_{HS}^2 = \sum_{i,j} |c_{ij}(A)|^2 \in [0, \infty]$$

is independent of the choice of orthonormal bases in H and H'.

PROOF By Parseval's theorem

$$\|A\|_{HS}^2 = \sum_{i,j} |c_{ij}(A)|^2 = \sum_{i} \|Ae_i\|^2$$

which is certainly independent of the choice of basis in H'. But since $c_{ij}(A) = \overline{c}_{ji}(A^*)$, $\|A\|_{HS}^2 = \|A^*\|_{HS}^2$ which is independent of the choice of basis in H by the same argument. □

DEFINITION 8.2 An operator A such that $\|A\|_{HS} < \infty$ is called a *Hilbert-Schmidt operator*, and $\|A\|_{HS}$ is called its *Hilbert-Schmidt norm*.

PROPOSITION 8.3

(i) *The Hilbert-Schmidt norm is induced by an inner product*

$$\langle A, B \rangle_{HS} = \sum_{i,j} \overline{c}_{ij}(A) c_{ij}(B) .$$

(ii) *Relative to this inner product, the space of Hilbert-Schmidt operators is a Hilbert space.*

(iii) *The Hilbert-Schmidt norm dominates the operator norm.*

---

[1] Of course, not every such infinite matrix represents a bounded operator; but this does not matter here.

(iv) *Hilbert-Schmidt operators are compact.*

(v) *The sum of two Hilbert-Schmidt operators, and the product (in either order) of a Hilbert-Schmidt and a bounded operator are Hilbert-Schmidt.*

The proofs are all easy. Our interest in the Hilbert-Schmidt operators is as a crutch to get us to trace-class operators, which we now define.

DEFINITION 8.4 A bounded operator T on a Hilbert space H is said to be of *trace-class* if there are Hilbert-Schmidt operators A and B on H with $T = AB$. Its *trace* $\text{Tr}(T)$ is defined to be the Hilbert-Schmidt inner product $\langle A^*, B \rangle_{HS}$.

A priori, the trace depends on the choice of A and B. However

$$\text{Tr}(T) = \sum_{i,j} \overline{c}_{ij}(A^*) c_{ij}(B) = \sum_{i,j} c_{ji}(A) c_{ij}(B) = \sum_j c_{jj}(T) \tag{8.5}$$

in fact depends only on T.

REMARK 8.6 We have now defined several classes of operators:

$$(\text{trace-class}) \subset (\text{Hilbert-Schmidt}) \subset (\text{compact}) \subset (\text{bounded}).$$

This sequence of inclusions should be thought of as the "non-commutative analogue" of the sequence of inclusions

$$l^1 \subset l^2 \subset c_0 \subset l^\infty$$

of sequence spaces, see Simon [68].

PROPOSITION 8.7 *Let T be self-adjoint and of trace class. Then $\text{Tr}(T)$ is the sum of the eigenvalues of T.*

PROOF Choose an orthonormal basis of eigenvectors (which exists by the spectral theorem for compact self-adjoint operators) and apply (8.5). □

The conclusion still holds if T is not self-adjoint, a result known as *Lidskii's theorem*. This is very much harder to prove.

The most important fact about the trace is its commutator property:

PROPOSITION 8.8 *Let T and B be bounded operators on a Hilbert space H, and suppose that either T is of trace-class, or both T and B are Hilbert-Schmidt. Then TB and BT are trace-class, and $\mathrm{Tr}(TB) = \mathrm{Tr}(BT)$.*

PROOF That TB and BT are trace-class follows from (8.4) and (8.3) v). Now choose an orthonormal basis (e_i) for H, and write

$$\begin{aligned}
\mathrm{Tr}(TB) &= \sum_i \langle TBe_i, e_i \rangle \\
&= \sum_i \langle Be_i, T^*e_i \rangle \\
&= \sum_{i,j} \bar{c}_{ij}(B) \bar{c}_{ij}(T) \text{ (by Parseval's theorem)}.
\end{aligned}$$

This sum is absolutely convergent, and it is symmetrical in B and T, so the result follows. □

Examples of Hilbert-Schmidt and trace-class operators come from integral operators on manifolds.

PROPOSITION 8.9 *Let M be a compact manifold equipped with a smooth volume form vol (e.g. an oriented Riemannian manifold) and let A be a bounded operator on $L^2(M)$ defined by*

$$Au(m_1) = \int_M k(m_1, m_2) u(m_2)\, \mathrm{vol}(m_2)$$

where k is continuous on $M \times M$. Then A is a Hilbert-Schmidt operator, and

$$\|A\|_{HS}^2 = \iint |k(m_1, m_2)|^2\, \mathrm{vol}(m_1)\, \mathrm{vol}(m_2) .$$

PROOF Choose an orthonormal basis (e_j) for $L^2(M)$, and recall from the proof of (8.1) that

$$\begin{aligned}
\|A\|_{HS}^2 &= \sum_j \|Ae_j\|^2 \\
&= \sum_j \int \left| \int k(m_1, m_2) e_j(m_2)\, \mathrm{vol}(m_2) \right|^2 \mathrm{vol}(m_1) \\
&= \int \sum_j \left| \int k(m_1, m_2) e_j(m_2)\, \mathrm{vol}(m_2) \right|^2 \mathrm{vol}(m_1) .
\end{aligned}$$

But by Parseval's theorem

$$\sum_j \left| \int k(m_1, m_2) e_j(m_2) \operatorname{vol}(m_2) \right|^2 = \int |k(m_1, m_2)|^2 \operatorname{vol}(m_2)$$

so

$$\|A\|_{HS}^2 = \iint |k(m_1, m_2)|^2 \operatorname{vol}(m_1) \operatorname{vol}(m_2) < \infty$$

as asserted. □

THEOREM 8.10 *Now let M and A be as in (8.9), but assume that k is smooth on $M \times M$, so that A is a smoothing operator. Then A is of trace-class, and*

$$\operatorname{Tr}(A) = \int k(m, m) \operatorname{vol}(m) \, .$$

PROOF Suppose first of all that $A = BC$, where B and C are Hilbert-Schmidt operators represented by continuous kernels k_B and k_C, as in (8.9). Then

$$k(m_1, m_3) = \int k_B(m_1, m_2) k_C(m_2, m_3) \operatorname{vol}(m_2) \, .$$

The trace of A is the Hilbert-Schmidt inner product of B^* and C. However, (8.9) determines the Hilbert-Schmidt norm, and therefore by polarization the Hilbert-Schmidt inner product also, on the space of operators with continuous kernels. Thus,

$$\begin{aligned} \operatorname{Tr}(A) &= \iint k_B(m_1, m_2) k_C(m_2, m_1) \operatorname{vol}(m_1) \operatorname{vol}(m_2) \\ &= \int k(m, m) \operatorname{vol}(m) \, . \end{aligned}$$

So all we need to check is that any smoothing operator A can be written in the form BC. However, by remark 5.32, the operator $(1 + \Delta)^{-N}$ for sufficiently large N is a Hilbert-Schmidt operator with continuous kernel. Thus we may write any smoothing operator A in the form BC, where $B = (1 + \Delta)^{-N}$ has continuous kernel and $C = (1 + \Delta)^{+N} A$ is a smoothing operator. □

REMARK 8.11 In (8.9) and (8.10) we dealt for simplicity only with operators on functions on M, which have scalar valued kernels. We will want to use the corresponding results which apply to operators on sections of a vector bundle S, where the kernels have values in $S \boxtimes S^*$. Then $k(m, m) \in S_m \otimes S_m^* \cong \operatorname{Hom}(S_m, S_m)$, and the statement corresponding to (8.10) is

THEOREM 8.12 *Let A be a smoothing operator on $L^2(S)$, with kernel k. Then A is of trace-class, and*

$$\operatorname{Tr}(A) = \int \operatorname{tr} k(m,m)\operatorname{vol}(m)$$

where $\operatorname{tr}\colon S_m \otimes S_m^* \to \mathbb{C}$ denotes the canonical trace on endomorphisms of the finite dimensional vector space S_m.

To prove this, use local trivializations and partitions of unity to reduce to (8.10).

Weyl's asymptotic formula

Recall the situation considered in the introduction to this chapter: M is a compact oriented Riemannian manifold, and the spectrum of the Laplace operator Δ on M is a discrete set of positive real numbers,

$$\lambda_0 \leqslant \lambda_1 \leqslant \lambda_2 \leqslant \cdots,$$

tending to infinity. One can vaguely think of these numbers as the "resonant frequencies" of M under some kind of "oscillation". It is then natural to ask to what extent the geometry of M can be recovered from this set of "resonant frequencies", a question put memorably in the title of Kac's paper [42].

Most approaches to this question rely on the following idea. The operator $e^{-t\Delta}$ is smoothing, hence of trace-class (8.10), and its trace is given by integration of its kernel over the diagonal. So, from the asymptotic expansion (7.15),

$$\operatorname{Tr}(e^{-t\Delta}) \sim \frac{1}{(4\pi t)^{n/2}}(a_0 + ta_1 + \cdots)$$

where

$$a_i = \int_M \Theta_i(m)\operatorname{vol}(m).$$

On the other hand, by (8.7),

$$\operatorname{Tr}(e^{-t\Delta}) = \sum_j e^{-t\lambda_j}.$$

Therefore

$$(4\pi t)^{n/2}\sum_j e^{-t\lambda_j} \sim a_0 + ta_1 + \cdots \tag{8.13}$$

and we see that the spectrum of Δ determines the quantities $n, a_0, a_1, \ldots$ which encode geometric information about the manifold M. The more coefficients we can calculate, the more we will know about spectral geometry.

We know that $a_0 = \text{vol}(M)$ and $a_1 = \frac{1}{6} \int_M \kappa(m) \text{vol}(m)$ by 7.19. Thus we can state:

PROPOSITION 8.14 *The spectrum of the Laplacian on M determines the dimension, the volume and the total scalar curvature of M. In case $\dim(M) = 2$, it determines the topology of M.*

PROOF The first statement follows from the expansion (8.13). As for the second, in dimension 2, the total curvature determines the topology, because of the Gauss-Bonnet theorem and the classification of 2-manifolds. □

REMARK 8.15 In more general contexts it may not be the case that the curvature determines the geometry, or indeed even the topology. For the resolution of Kac's question in the original context of planar domains see [35].

The formula (8.13) can also be used in the other direction; knowing (some of) the ϑ's, we try to discover information about the spectrum. We will prove a famous theorem of Weyl in this direction.

THEOREM 8.16 *Let $\mathfrak{N}(\lambda)$ denote the number of eigenvalues less than λ. Then as $\lambda \to \infty$,*

$$\mathfrak{N}(\lambda) \sim \frac{1}{(4\pi)^{n/2}\Gamma((n/2)+1)} \text{vol}(M)\lambda^{n/2}.$$

REMARK 8.17 This theorem may be reformulated as an asymptotic estimate for the j'th eigenvalue:

$$\lambda_j \sim 4\pi \left(\frac{1}{\Gamma((n/2)+1)} \text{vol}(M) \right)^{2/n} j^{2/n}.$$

PROOF We can think of the sum $\sum e^{-t\lambda_j}$ as giving us an "exponentially weighted average count" of the number of eigenvalues of Δ. To reconstruct $\mathfrak{N}$ from such averages is the task of *Tauberian theory*. In fact, from 8.13 we have

$$t^\alpha \sum e^{-t\lambda_j} \to A \qquad (*)$$

as $t \to 0$, where $\alpha = n/2$ and $A = (4\pi)^{-n/2}\operatorname{vol}(M)$. An abstract Tauberian theorem of Karamata states that for any nondecreasing sequence of positive numbers λ_j having the convergence property $(*)$ for some $A, \alpha > 0$, the associated counting function $\mathfrak{N}(\lambda)$ satisfies $\mathfrak{N}(\lambda) \sim A\lambda^\alpha/\Gamma(\alpha+1)$ as $\lambda \to \infty$. This immediately implies Weyl's theorem, of course; so it remains for us to prove Karamata's result.

For any continuous function f on $[0,1]$, let us define

$$\varphi_f(t) = \sum f(e^{-t\lambda_j})e^{-t\lambda_j}.$$

I claim that, for every f,

$$t^\alpha \varphi_f(t) \to \frac{A}{\Gamma(\alpha)}\int_0^\infty f(e^{-s})s^{\alpha-1}e^{-s}\,ds \tag{8.18}$$

as $t \to 0$. An application of the Stone-Weierstrass theorem shows that it is enough to prove this result when f is a monomial of the form $f(x) = x^n$. But then the left hand side of 8.18 is $t^\alpha \sum e^{-(n+1)t\lambda_j}$, which tends to $A(n+1)^{-\alpha}$ as $t \to 0$; and direct calculation shows that the right hand side is equal to $A(n+1)^{-\alpha}$ also.

For $r < 1$ let $f_r \colon [0,1] \to \mathbb{R}$ be the continuous function such that $f(x) = 0$ for $x \in [0, r/e]$, $f(x) = 1/x$ for $x \in [1/e, 1]$, and $f(x)$ linearly interpolates between 0 and $1/e$ on the interval $[r/e, 1/e]$. We apply 8.18 to f_r, and put $t = 1/\lambda$. Notice that

$$\varphi_{f_r}(1/r\lambda) \leqslant \mathfrak{N}(\lambda) \leqslant \varphi_{f_r}(1/\lambda).$$

Let $\lambda \to \infty$ and make easy estimates to obtain

$$\limsup \lambda^{-\alpha}\mathfrak{N}(\lambda) \leqslant \frac{A}{\alpha\Gamma(\alpha)}, \quad \liminf \lambda^{-\alpha}\mathfrak{N}(\lambda) \geqslant \frac{Ar^\alpha}{\alpha\Gamma(\alpha)}.$$

Now r is arbitrary, so $\lim \lambda^{-\alpha}\mathfrak{N}(\lambda)$ exists and equals $A/\alpha\Gamma(\alpha)$. Bearing in mind the standard identity $\alpha\Gamma(\alpha) = \Gamma(\alpha+1)$, we obtain Karamata's theorem. □

REMARK 8.19 Though we have worked all through this chapter with the scalar Laplacian Δ, which is the classical case, it is clear that the methods all work to count the eigenvalues of the general Laplace-type operator D^2. Then Θ_0 becomes the identity endomorphism of the Clifford bundle S, and tracing this according to 8.1 yields an additional factor $\dim(S)$ in the formulae, but otherwise everything is the same.

Notes

A classic reference for trace-class operators and related 'operator ideals' is Simon [68].

Further information about spectral geometry may be found in the book by Berger et al [11]. As well as the kind of analysis carried out in this chapter, which relates to the asymptotic behavior of the large eigenvalues, one can investigate the lowest eigenvalues of the Laplacian and relate them to geometric properties such as isoperimetric constants.

In the first edition of this book we derived Karamata's Tauberian theorem from the general method of Wiener, expounded in many books on functional analysis [64]. The simple direct proof given here is borrowed from [12].

For more about isospectral manifolds, see Brooks [16].

Exercises

QUESTION 8.20 Prove that the sum of two trace-class operators is trace-class.

QUESTION 8.21 Give an example of an operator A defined by a continuous kernel k, that is

$$Au(x) = \int k(x,y)u(y)\,dy$$

which is *not* of trace-class. (Hint: Consider the convergence of Fourier series.) Prove however that *if A is* of trace class, then

$$\mathrm{Tr}(A) = \int k(x,x)\,dx.$$

QUESTION 8.22 Prove the following quantitative version of Rellich's theorem: the inclusion of the Sobolev space $W^{k+\ell}(M)$ into $W^k(M)$ is a Hilbert-Schmidt operator for $\ell > \frac{1}{2}\dim(M)$. Deduce another proof that smoothing operators are of trace class.

QUESTION 8.23 Let λ_j be the eigenvalues of the Laplacian Δ of a compact manifold, arranged in increasing order. Define the *zeta-function* by

$$\zeta(s) = \sum \lambda_j^{-s},$$

the summation being taken over the non-zero eigenvalues. By (8.16), this Dirichlet series converges for $\Re(s) \gg 0$. Show that in fact $\zeta(s)$ extends to a meromorphic function of s on the entire complex plane, and that

$$\zeta(0) = \frac{1}{(4\pi)^{n/2}} \int \Theta_{n/2}(m) \, \text{vol}(m)$$

where $\Theta_{n/2}$ is the asymptotic-expansion coefficient as defined in (7.15). (Use the formula $\Gamma(s)\lambda^{-s} = \int_0^\infty e^{-t\lambda} t^{s-1} \, dt$.)

CHAPTER 9

Some non-compact manifolds

In this chapter we look at some Dirac-type operators on non-compact manifolds. The example of the operator id/dx on the real line, as compared to the corresponding operator on the circle $\mathbb{R}/\mathbb{Z}$, is a helpful one to bear in mind. On the circle, id/dx has a discrete spectrum with finite-dimensional eigenspaces. On the line, Fourier series are replaced by Fourier transforms; the spectrum becomes continuous, and spectral values no longer correspond to square-integrable eigenfunctions. Nevertheless, a spectral decomposition still exists, and the functional calculus operates in the same way as before.

From the perspective of quantum mechanics (see for example [**65**, Chapters 4,5]) the appearance of continuous spectrum is related to the presence of non-localized or 'unbound' solutions for the corresponding Schrödinger equation. Now, if one adds to the Schrödinger equation a term representing a 'potential well', then solutions of the equation representing energies lower than the depth of the well will be 'bound' within the well. We therefore expect that the discrete spectrum property can be restored by adding to the Dirac-type operator a potential term sufficiently strong to localize all the eigenfunctions. In the first section of this chapter we will discuss a famous example of this phenomenon, the *quantum harmonic oscillator*. In subsequent sections we will show how operators of this sort can arise geometrically, and we will give some of the general theory in the absence of a localizing potential.

Only the first section of this chapter is required for our proof of the Index Theorem; the remaining sections are needed only for the more specialized applications at the end of the book.

The harmonic oscillator

DEFINITION 9.1 The *harmonic oscillator* is the name given to the unbounded

operator
$$H = -\frac{d^2}{dx^2} + a^2 x^2 \quad (a > 0)$$
on $L^2(\mathbb{R})$.

For the physical meaning of this operator see [65, §13]. Its relevance to index theory was observed by E. Getzler. For the moment, we will work out a few facts about the spectral theory of H.

DEFINITION 9.2 The *annihilation operator* A is defined by $A = ax + d/dx$, the *creation operator* A^* is $A^* = ax - d/dx$.

The operators A, A^*, H may be taken to have as domain the Schwartz space $\mathcal{S}(\mathbb{R})$, and they map $\mathcal{S}(\mathbb{R})$ to $\mathcal{S}(\mathbb{R})$. Elementary computations give

$$\left. \begin{array}{rclrcl} AA^* &=& H + a, & A^*A &=& H - a \\ [A, A^*] &=& 2a & & & \\ [H, A] &=& -2aA, & [H, A^*] &=& 2aA^* \end{array} \right\} \quad (9.3)$$

DEFINITION 9.4 The *ground state* of H is the function $\psi_0 \in L^2(\mathbb{R})$ satisfying the differential equation $A\psi_0 = 0$ and such that $\|\psi_0\| = 1$.

Clearly, then, $H\psi_0 = a\psi_0$, so ψ_0 is an eigenfunction of H. In fact, we can calculate ψ_0 explicitly, thus checking that it is square-integrable:

$$\frac{d\psi_0}{dx} + ax\psi_0 = 0 .$$

Therefore
$$\int \frac{d\psi_0}{\psi_0} = -a \int x \, dx .$$

The solution is $\psi_0 = Ce^{-ax^2/2}$, where C is a normalizing constant; since $\|\psi_0\| = 1$ $C = a^{\frac{1}{2}} \pi^{\frac{1}{4}}$.

Now for $k \geqslant 1$, define the *excited states* of H inductively by

$$\psi_k = \frac{1}{(2ka)^{\frac{1}{2}}} A^* \psi_{k-1} . \quad (9.5)$$

LEMMA 9.6 ψ_k belongs to the Schwartz class. It is a normalized eigenfunction of H, with eigenvalue $(2k+1)a$.

PROOF Induction. We have

$$\begin{aligned}
H\psi_k &= \frac{1}{(2ka)^{\frac{1}{2}}} H A^* \psi_{k-1} \\
&= \frac{1}{(2ka)^{\frac{1}{2}}} (A^* H + 2aA^*) \psi_{k-1} \quad \text{(by (9.3))} \\
&= \frac{1}{(2ka)^{\frac{1}{2}}} A^*((2k-1)a\psi_{k-1} + 2a\psi_{k-1}) \\
&= (2k+1)a\psi_k \ .
\end{aligned}$$

Similarly

$$\begin{aligned}
\|\psi_k\|^2 &= \frac{1}{2ka} \langle A^* \psi_{k-1}, A^* \psi_{k-1} \rangle \\
&= \frac{1}{2ka} \langle A A^* \psi_{k-1}, \psi_{k-1} \rangle \\
&= \frac{1}{2ka} \langle (H+a)\psi_{k-1}, \psi_{k-1} \rangle \\
&= \frac{1}{2ka} \langle (2ka\psi_{k-1}), \psi_{k-1} \rangle \\
&= 1 \quad \square .
\end{aligned}$$

LEMMA 9.7 $\psi_k(x) = h_k(x)e^{-ax^2/2}$, where h_k is a polynomial of degree k with positive leading coefficient.

PROOF Induction, using the recurrence relation

$$h_k(x) = \frac{1}{(2ka)^{\frac{1}{2}}}((a+1)x h_{k-1}(x) - h'_{k-1}(x))$$

which follows easily from (9.5). $\square$

Up to a normalization, the h_k's are the well known Hermite polynomials. From 9.7, it follows that the linear span of the $\{\psi_k\}$ is the space

$$\mathcal{P} = \{x \mapsto p(x)e^{-ax^2/2} : p \text{ polynomial}\} \leqslant L^2(\mathbb{R}) \ .$$

PROPOSITION 9.8 $\mathcal{P}$ is dense in $L^2(\mathbb{R})$.

PROOF We may assume $a = 1$. Let $f_j(x) = x^j e^{-x^2/2}$. We calculate

$$\begin{aligned}
\|f_j\|^2 &= \int_{-\infty}^{\infty} x^{2j} e^{-x^2} \, dx \\
&= 2 \int_0^{\infty} y^j e^{-y} \frac{dy}{2\sqrt{y}} \quad \text{on substituting } y = x^2 \\
&= \Gamma(j + \tfrac{1}{2}) \leqslant j!
\end{aligned}$$

Now we may write

$$e^{i\lambda x - x^2/2} = \sum_{j=0}^{\infty} \frac{(i\lambda)^j}{j!} f_j(x) \, .$$

The L^2 norms of the terms in this series are bounded by $|\lambda|^j (j!)^{\frac{1}{2}}$, so it is convergent in L^2. Since $f_j \in \mathcal{P}$, the functions $x \mapsto e^{i\lambda x - x^2/2}$ belong to $\overline{\mathcal{P}}$. Now suppose that $f \in L^2$ is orthogonal to $\overline{\mathcal{P}}$; then

$$\int_{-\infty}^{\infty} f(x) e^{i\lambda x - x^2/2} \, dx = 0 \quad \forall \lambda \in \mathbb{R} \, .$$

But then by Plancherel's theorem $f(x) e^{-x^2/2} = 0$ almost everywhere, so $f = 0$ almost everywhere. □

We have shown that the space $L^2(\mathbb{R})$ admits a complete orthogonal decomposition into (1-dimensional) eigenspaces for H, with discrete spectrum tending to infinity. This is exactly the conclusion of Theorem 5.27 for the Dirac operator. We see therefore that spectrally H is like the Dirac operator on a compact manifold. In physical language, the states of the harmonic oscillator are all 'bound states'. We may if we wish define the analogs of the Sobolev spaces, making use of the eigenfunctions ψ_k instead of the characters e^{-imx} on the torus. We will need only one result of this kind:

LEMMA 9.9 *Let $u \in L^2(\mathbb{R})$. Then $u \in \mathcal{S}(\mathbb{R})$ if and only if the "Fourier coefficients" $a_k = \langle \psi_k, u \rangle$ are rapidly decreasing in k.*

PROOF If $u \in \mathcal{S}(\mathbb{R})$, then for all l, $H^l u \in \mathcal{S}(\mathbb{R})$, and since H^l acts on the Fourier coefficients by multiplying a_k by $((2k+1)a)^l$, the result is obvious.

Conversely, suppose that the Fourier coefficients are rapidly decreasing. Then for all l, $(A^*)^l u$ has rapidly decreasing Fourier coefficients, by (9.5), and so does $A^l u$; so $D^l u$ and $M^l u$ have rapidly decreasing Fourier coefficients where D and M

denote the differentiation and multiplication operators id/dx and x. Therefore, for any noncommutative polynomial p, $p(M, D)u \in L^2(\mathbb{R})$; and it is easy to check that this implies $u \in \mathcal{S}(\mathbb{R})$. □

PROPOSITION 9.10 *If f is a bounded function on the spectrum of H, then $f(H)$ is defined and is a bounded operator on $L^2(\mathbb{R})$; the map $f \mapsto f(H)$ is a homomorphism from the ring of bounded functions on the spectrum of H to $B(L^2(\mathbb{R}))$. Moreover, $f(H)$ maps $\mathcal{S}(\mathbb{R})$ to $\mathcal{S}(\mathbb{R})$.*

PROOF As 5.30, but using 9.9 instead of the Sobolev embedding theorem. □

We now have enough analysis to carry over much of the earlier discussion of the heat equation to the 'harmonic oscillator heat equation'

$$\frac{\partial u}{\partial t} + Hu = 0. \tag{9.11}$$

Indeed, the solution operator e^{-tH} is defined by the Hilbert space theory above (9.10); and there is a heat kernel $k_t^H \in \mathcal{S}(\mathbb{R} \times \mathbb{R})$ such that

$$e^{-tH}u(x) = \int k_t^H(x,y)u(y)\,dy.$$

As in the compact manifold case, the heat kernel is characterized by the facts that it belongs to $\mathcal{S}$, satisfies the heat equation in the x-variable, and tends to a δ-function as $t \to 0$.

We will need an explicit expression for the heat kernel $k_t^H(x, y)$, at least when $y = 0$ (see exercise 9.23) for the general case. To get one, proceed by inspired guesswork. Try

$$k_t^H(x, 0) = u(x, t) = \alpha(t)e^{-\frac{1}{2}\beta(t)x^2}$$

where α and β are functions to be determined[1]. Then

$$Hu = \alpha(x^2(a^2 - \beta^2) + \beta)e^{-\frac{1}{2}\beta(t)x^2}$$
$$\frac{\partial u}{\partial t} = -(x^2\dot{\beta}\alpha/2 + \dot{\alpha})e^{-\frac{1}{2}\beta(t)x^2}.$$

Summing and equating coefficients to zero, we get

$$\dot{\beta} = 2(a^2 - \beta^2)\,,\ \dot{\alpha} = -\beta\alpha\,.$$

---
[1] A trial solution like this is sometimes referred to as an 'ansatz'.

We solve these differential equations, looking for a solution with $\beta(t) \to \infty$ like $1/t$ as $t \to 0$. This gives

$$\beta(t) = a\coth(2at)$$
$$\alpha(t) = (\text{const.})(\sinh 2at)^{-\frac{1}{2}}.$$

If we choose the constant to equal $(a/2\pi)^{\frac{1}{2}}$, then as $t \to 0$,

$$u(x,t) \sim \frac{1}{\sqrt{4\pi t}} e^{-x^2/4t}$$

and we know that for any function $s \in \mathcal{S}(\mathbb{R})$,

$$\frac{1}{\sqrt{4\pi t}} \int e^{-x^2/4t} s(x)\, dx \to s(0) \quad \text{as } t \to 0.$$

Thus $u(x,t)$ satisfies the heat equation and tends to a δ-function, so is equal to the heat kernel. We state this result more formally:

PROPOSITION 9.12 *The harmonic oscillator heat kernel satisfies*

$$u(x,t) = \sqrt{\frac{a}{2\pi \sinh(2at)}} \exp\left(\frac{-ax^2 \coth(2at)}{2}\right).$$

This result is known as *Mehler's formula*.

REMARK 9.13 Our analysis of the harmonic oscillator requires the assumption that a is a *real* number. However, the function $u(x,t)$ described in 9.12 above is clearly analytic as a function of a for $a \in \mathbb{C}$, $|a| < \pi/2t$. By analytic continuation, then, we find that $u(x,t)$ continues to satisfy the equation

$$\frac{\partial u}{\partial t} - \frac{\partial^2 u}{\partial x^2} + a^2 x^2 u = 0$$

even if a is a (sufficiently small) *complex* number. This analytic continuation will be of importance in the proof of the index theorem. Of course, if a is not real then may not be in the Schwarz class, and our uniqueness theorems do not apply directly

Witten's perturbation of the de Rham complex

Let M be a complete Riemannian manifold (compact or not) and let $h\colon M \to \mathbb{R}$ be a smooth function. In [73], E. Witten introduced a certain perturbation of the de Rham complex of M, determined by the function h; and this perturbation has become important in several contexts. In particular, variants of the harmonic oscillator appear naturally as the Laplacians of perturbed de Rham complexes on Euclidean space.

DEFINITION 9.14 Let M, h be as above. The *perturbed exterior derivative* d_s (depending on the parameter $s \in \mathbb{R}^+$) is defined by

$$d_s\omega = e^{-sh}d(e^{sh}\omega) = d\omega + sdh \wedge \omega.$$

Its adjoint is given by

$$d_s^*\omega = e^{sh}d^*(e^{-sh}\omega) = d^*\omega - sdh \lrcorner\, \omega$$

Note that this agrees with the calculation of 3.21, that interior multiplication is minus the adjoint of exterior multiplication. The perturbed analogue of the de Rham operator is

$$D_s = d_s + d_s^* = D + sR \qquad (9.15)$$

where R is the endomorphism of the exterior bundle given by $(dh\wedge) - (dh\lrcorner)$.

This formula is most conveniently expressed in terms of the Clifford bimodule structure of the exterior algebra. Recall that $\bigwedge^* T^*M$ is isomorphic to the Clifford algebra itself, and therefore carries both a left and a right multiplication action of the Clifford algebra; these actions commute. For $e \in TM$ let us define

$$L_e\omega = e \cdot \omega, \qquad R_e\omega = (-1)^{\partial\omega}\omega \cdot e.$$

Here the dot denotes Clifford multiplication and $\partial\omega = p$ for $\omega \in \bigwedge^p$. Notice that, because of the extra sign, L_e and $R_{e'}$ now anticommute for any e, e'. The endomorphism R of 9.15 is equal to $R_{\nabla h}$ in this new notation. Equivalently, in terms of a local orthonormal frame e_i, we may write formula 9.15 as

$$D_s\omega = \sum_i \left(L_{e_i}\nabla_i\omega + s(e_i \cdot h)R_{e_i}\omega\right)$$

where the dot denotes the Lie derivative.

We will need a Weitzenbock-type formula for D_s^2.

DEFINITION 9.16 Let $x \in M$. The *Hessian* of h at x is the symmetric bilinear form H_h on $T_x M$ defined by
$$H_h(X,Y) = X \cdot (Y \cdot f)(x) - (\nabla_X Y) \cdot f(x).$$

It is easy to see that this formula is tensorial, that is, it depends only on the values of the vector fields X and Y at the point x. The Hessian is the bilinear form corresponding to the symmetric matrix of second derivatives of h (relative to a synchronous orthonormal frame at the point x). Let H_h be the endomorphism of $\wedge^* T^* M$ defined by
$$\mathsf{H}_h = \sum_{i,j} H_h(e_i, e_j) L_{e_i} R_{e_j}$$
relative to an orthonormal frame e_i; it is easy to check that this expression is independent of the choice of frame.

LEMMA 9.17 *With the notations of 9.15 above,*
 (i) *R^2 is the endomorphism given by multiplication by $|dh|^2$;*
 (ii) *$RD + DR = \mathsf{H}_h$ as an endomorphism of the exterior bundle.*
Consequently, we have the formula $D_s^2 = D^2 + s^2|dh|^2 + s\mathsf{H}_h$.

PROOF From the local coordinate formula for D_s we have
$$D_s^2 = \sum_{i,j} \Big(L_{e_i}\nabla_i + sR_{e_i}(e_i \cdot h)\Big)\Big(L_{e_j}\nabla_j + sR_{e_j}(e_j \cdot h)\Big)$$
and a direct calculation, remembering that L and R anticommute and that $R_{e_i} R_{e_j} = \delta_{ij}$, gives the result. $\square$

We consider the special case of Euclidean space. Let $M = \mathbb{R}^n$ with its standard metric and let $h(x) = \frac{1}{2}\sum \lambda_j x_j^2$, a quadratic form on M. The lemma above gives in this case
$$D_s^2 = \sum_j \Big(-\Big(\frac{\partial}{\partial x^j}\Big)^2 + s^2(\lambda_j x^j)^2 + s\lambda_j Z_j\Big)$$
where $Z_j = [dx^j \lrcorner, dx^j \wedge]$ is the operator which is equal to $+1$ on a basis element $dx^{i_1} \wedge \cdots \wedge dx^{i_k}$ which has $j \in \{i_1, \ldots, i_k\}$, and equal to -1 on such a basis element

which has $j \notin \{i_1, \ldots, i_k\}$. Notice that the first two terms in the sum are precisely the harmonic oscillator, discussed at the beginning of the chapter.

PROPOSITION 9.18 *Consider $M = \mathbb{R}^n$ equipped with the quadratic function h described above. Then, for $s > 0$, there is a basis for $L^2(\mathbb{R}^n)$ consisting of smooth, rapidly decaying eigenfunctions for the operator D_s^2; the corresponding eigenvalues are the numbers*

$$s \sum_j \Big(|\lambda_j|(1 + 2p_j) + \lambda_j q_j \Big),$$

where $p_j = 0, 1, 2, \ldots$ and $q_j = \pm 1$. If we consider the action of D_s^2 on k-forms, the spectrum is as above with the additional restriction that precisely k of the numbers q_j are equal to $+1$.

PROOF Let us write $Y_j = -\left(\dfrac{\partial}{\partial x^j}\right)^2 + s^2 \lambda_j^2 (x^j)^2$, so that Y_j is a harmonic oscillator in the j-variable. The Z and Y operators all commute, so they can be simultaneously diagonalized. By the spectral theory of the harmonic oscillator, we know that $\sum Y_j$ is essentially self-adjoint, with discrete spectrum: its eigenvalues are the numbers $\sum |\lambda_j|(1+2p_j)$, and each of these eigenvalues has multiplicity 2^n (the fiber dimension of the exterior bundle). The operators Z_j act on each of the eigenspaces as involutions, splitting them into ± 1 eigenspaces for each Z_j: the eigenspace with eigenvalue $\sum(|\lambda_j|(1+2p_j) + \lambda_j q_j)$ for L is precisely the q_j-eigenspace for each Z_j acting on the $\sum |\lambda_j|(1+2p_j)$-eigenspace for $\sum Y_j$. $\square$

Functional calculus on open manifolds

Let M be a complete Riemannian manifold, D a Dirac operator on a Clifford bundle S over M. (Our results will also be valid for a self-adjoint generalized Dirac operator of the form $D + A$.) In this section we will develop a functional calculus for D, that is, a ring-homomorphism $f \mapsto f(D)$ having properties analogous to those in 5.30 in the compact case.

The classical approach to these questions would require us to consider D as an unbounded operator on the Hilbert space $L^2(S)$, with domain $C_c^\infty(S)$. The operator D is formally self-adjoint on this domain, and one can use the completeness of M to prove that D is in fact *essentially self-adjoint* in the sense of unbounded

127

operator theory — meaning that the closure of D is equal to its Hilbert space adjoint, or equivalently that D has a unique self-adjoint extension. One then applies the spectral theorem for unbounded self-adjoint operators [29, Chapter XII] to produce the desired functional calculus.

We will take a different approach, which is motivated by an article of Chernoff [19]. Chernoff showed that one could use the finite propagation speed of solutions to the Dirac wave equation (7.20) to prove essential self-adjointness. We will use finite propagation speed to construct the functional calculus directly.

PROPOSITION 9.19 *The wave equation*

$$\frac{\partial s_t}{\partial t} = iDs$$

has a unique solution for smooth, compactly supported initial data s_0 on M; and the solution s_t is smooth and compactly supported for all times t.

PROOF Uniqueness follows from an energy estimate as in 7.4. For existence, suppose that we want to construct the solution s_t for all $|t| \leq t_0$, and suppose that $\mathrm{Supp}(s_0) = K \subseteq M$. We can build a *compact* manifold M' and a Clifford bundle S' over it, such that M' contains an open subset isometric to a t_0-neighbourhood U of K in M, by an isometry which is covered by an isomorphism of Clifford bundles.[2] Now the wave equation can be solved on M' by the results of Chapter 7, and finite propagation speed shows that for $|t| < t_0$ the support of the solution s_t remains within U; so s_t can be considered to be a solution on M as well. □

Notice that the argument above also shows that the wave equation on M has unique propagation speed. Notice also that since the solution operator e^{itD} is defined and unitary on a dense subspace (namely $C_c^\infty(S)$) of $L^2(S)$, it extends by continuity to a unitary operator on the whole of $L^2(S)$.

---

[2]This can be achieved, for example, by the 'doubling construction'; let $(M_0, \partial M_0)$ be a compact codimension-0 submanifold with boundary of M, such that the interior of M_0 contains a t_0-neighbourhood of K, and define M' by gluing two copies of M_0, one with orientation reversed, together along the boundary ∂M_0.

Now let $f \in \mathcal{S}(\mathbb{R})$ be a Schwarz-class function. Define the operator $f(D)$ by the Fourier integral

$$f(D) = \frac{1}{2\pi} \int \hat{f}(t) e^{itD} \, dt$$

where e^{itD} is the unitary solution operator to the wave equation, described above. Since $\hat{f}$ is of rapid decay, the integral defining $f(D)$ does converge in the weak sense of 7.22.

PROPOSITION 9.20 *The mapping $f \mapsto f(D)$ is a ring homomorphism from $\mathcal{S}(\mathbb{R})$ to $B(L^2(S))$. Moreover,*

$$\|f(D)\| \leqslant \sup |f|.$$

If $f(x) = x g(x)$, then $f(D) = D g(D)$.

PROOF All three parts of this theorem are proved in the same way, that is, by reduction to the case of compact manifolds. We give the proof of the second part and leave the others to the reader. Let $a = \sup|f|$. Suppose first of all that f has compactly supported Fourier transform, say that $\mathrm{Supp}(\hat{f}) \subset [-t_0, t_0]$. Let s be a compactly supported section of S, say with $\mathrm{Supp}(s) = K$. As in the previous proof, construct a compact manifold M' isometric to M on a t_0-neighbourhood U of K. Then by finite propagation speed, $f(D)s$ is supported within U and agrees with $f(D')s$, where D' is the Dirac operator on M'. But $\|f(D')s\| \leqslant a\|s\|$ by the functional calculus on the compact manifold M'. Thus we have shown that

$$\|f(D)s\| \leqslant a\|s\|$$

for all compactly supported sections s; hence for all $s \in L^2(S)$ be a density argument. Finally this inequality holds for all $f \in \mathcal{S}(\mathbb{R})$, since functions with compactly supported Fourier transform are dense in $\mathcal{S}(\mathbb{R})$. □

REMARK 9.21 Since the mapping $f \mapsto f(D)$ has $\|f(D)\| \leqslant \sup|f|$, it can be extended by continuity to a map from $C_0(\mathbb{R})$ to $B(L^2(S))$ having the same properties. Here $C_0(\mathbb{R})$ denotes the space of continuous functions on $\mathbb{R}$ vanishing at infinity; it is the sup norm closure of $\mathcal{S}(\mathbb{R})$.

Notes

The harmonic oscillator is discussed in many physics texts, for example [65, 74].

Witten's perturbed de Rham complex was introduced in [73]. By considering the asymptotics of the complex as s becomes large, he was able to give an entirely novel approach to classical results of Morse and Smale relating the topology of M to the critical values of the function h (that is, the zeros of dh). We will discuss some of this theory in Chapter 14.

I owe question 9.24 to my colleague David Acheson.

Exercises

QUESTION 9.22 Let $h_k(x)$ be the polynomials defined in 9.7, where we assume for simplicity that $a = 1$. Prove that

$$\exp(2tx - t^2) = \pi^{1/4} \sum_{k=0}^{\infty} \sqrt{\frac{2^k}{k!}} h_k(x) t^k.$$

QUESTION 9.23 Use the ansatz $k_t^H(x,y) = \alpha(t)\exp(-\frac{1}{2}\beta(t)(x^2+y^2) - \gamma(t)xy)$ to derive the general version of Mehler's formula,

$$k_t^H(x,y) = \sqrt{\frac{a}{2\pi \sinh(2at)}} \exp\left(\frac{-a(x^2+y^2)\coth(2at) + 2\operatorname{cosech}(2at)xy}{2}\right)$$

QUESTION 9.24 Derive Mehler's formula from the more general ansatz

$$u = h(t)f(\eta), \quad \eta = \frac{x}{g(t)}$$

by separation of variables. Obtain also another solution to the heat equation of the form

$$u(x,t) = \frac{Cx}{(\sinh 2at)^{\frac{3}{2}}} \exp-\{\frac{ax^2}{2\tanh 2at}\}.$$

Why does not the existence of this second solution contradict the uniqueness of the heat kernel?

QUESTION 9.25 (Donnelly-Xavier [28]) Let M be a complete Riemannian manifold, h a smooth function on M with $|\nabla h| \leq 1$ everywhere. By using the identity $DR_{\nabla h} + R_{\nabla h}D = H_h$, prove that for any k-form ω,

$$2\|\omega\|\|D\omega\| \geq \int_M \sigma_k \omega \wedge *\omega$$

where the function σ_k is defined in terms of the eigenvalues λ_i of the Hessian of h by

$$\sigma_k(x) = \sum_{i=1}^n \lambda_i(x) - 2k \max\{\lambda_i(x) : i = 1, \ldots, n\}.$$

Consider the case where M is hyperbolic n-space, of constant curvature -1. Take h to be the distance from a point q far from the support of ω. Show that if $k < (n-1)/2$ then there is a positive constant C_k such that $\|D\omega\| > C_k\|\omega\|$. Deduce that the spectrum of the Laplacian on k-forms does not contain zero in this case.

QUESTION 9.26 Let M be a complete Riemannian manifold. Construct a sequence φ_n of smooth, compactly supported functions $M \to [0, 1]$ with $\bigcup \operatorname{supp} \varphi_n = M$, $\varphi_n = 1$ on $\operatorname{supp} \varphi_{n-1}$, and $|\nabla \varphi_n| \leq 1/n$ everywhere.

Now let D be a Dirac operator on M, and suppose that s belongs to the domain of the Hilbert space adjoint D^*. Show that, if we define $s_n = \varphi_n s$, then each s_n belongs to the domain of the closure of D, that $s_n \to s$ in L^2, and that $Ds_n = D^*s_n \to D^*s$ in L^2 also. Deduce that D is essentially self-adjoint.

CHAPTER 10

The Lefschetz formula

In this chapter we will find our first example of a topological invariant defined by elliptic operators. The topological invariance will come from a pairwise cancelation of eigenspaces, which is sometimes called "supersymmetry".

Lefschetz numbers

Let M be a manifold and $\varphi\colon M \to M$ a map. Then φ induces an endomorphism φ^* of the (complex) cohomology of M, and the *Lefschetz number* $L(\varphi)$ of φ is defined by

$$L(\varphi) = \sum_q (-1)^q \operatorname{tr}(\varphi^* \text{ on } H^q(M)). \tag{10.1}$$

The classical *Lefschetz formula* (see Spanier [69]) expresses $L(\varphi)$ as a sum over the fixed points of φ. In particular, if $L(\varphi) \neq 0$, then φ has got some fixed points!

We want to approach the Lefschetz formula analytically and we will do so in the more general context of Dirac complexes (6.1). Thus, let M be a compact oriented n-dimensional Riemannian manifold, and let (S, d) be a Dirac complex over M. Let φ be a smooth map from M to M. Then φ induces $\varphi^*\colon C^\infty(S) \to C^\infty(\varphi^*S)$. In case S is the de Rham complex, there is a natural bundle map $\zeta = \Lambda^* T^* \varphi$ from φ^*S to S, but for a general Dirac complex there is no such map, and we must assume the existence of a bundle map $\zeta\colon \varphi^*S \to S$ as part of our data. Thus there is a composite map

$$F = \zeta\varphi^* \colon C^\infty(S) \to C^\infty(S).$$

DEFINITION 10.2 If F (as above) is a map of complexes (i.e. $Fd = dF$), one says that (ζ, φ) is a *geometric endomorphism* of the given Dirac complex. Its *Lefschetz number* $L(\zeta, \varphi)$ is defined by

$$L(\zeta, \varphi) = \sum_q (-1)^q \operatorname{tr}(F^* \text{ on } H^q(S)).$$

This definition is arranged so that a smooth map φ induces a natural geometric endomorphism of the de Rham complex, and its Lefschetz number as defined by (10.2) agrees with the classical definition of (10.1).

To apply analysis to the calculation of the Lefschetz number, we use the Hodge theorem (6.2); $H^q(S)$ is represented by the space $\mathcal{H}^q$ of harmonic sections of S_q. So if we define P_q to be the orthogonal projection $L^2(S_q) \to \mathcal{H}^q$, then

$$\operatorname{tr}(F^* \text{ on } H^q(S)) = \operatorname{Tr}(FP_q) \qquad (10.3)$$

where F is considered as a continuous linear operator from $C^\infty(S_q)$ to $L^2(S_q)$.

REMARK 10.4 Beware that the operator F itself may not be bounded on L^2; composition with φ can increase the L^2 norm of a smooth function by an arbitrarily large amount if φ happens to be constant on a nonempty open set. But it is apparent that F is bounded from $C^0(S_q)$ (the space of continuous sections) to $L^2(S_q)$, and therefore that the composite of F with any smoothing operator is bounded on L^2. This suffices for the arguments that follow.

LEMMA 10.5 *The operators P_q are smoothing operators. Moreover, if Δ_q denotes the restriction of D^2 to $C^\infty(S_q)$, then as $t \to \infty$ the smoothing kernel of $e^{-t\Delta_q}$ tends to the smoothing kernel of P_q in the C^∞ topology.*

PROOF P_q can be written as $f(\Delta_q)$, where $f(0) = 1$ and $f(\lambda) = 0$ for all other λ. Unfortunately, this function is not smooth. However, since Δ_q has discrete spectrum, there is a smooth function f of compact support equal to 1 at zero and equal to 0 at all other eigenvalues of Δ_q. Then $f \in \mathcal{S}(\mathbb{R})$ and $f(\Delta_q) = P_q$. If $g_t(x) = (1 - f(x))e^{-tx}$ then $g_t(\Delta_q) = e^{-t\Delta_q} - P_q$ and $g_t \to 0$ in $\mathcal{S}(\mathbb{R})$ as $t \to \infty$. Hence, by the functional calculus, $g_t(\Delta_q) \to 0$, and the result follows. □

Therefore $\operatorname{Tr}(FP_q) = \lim_{t \to \infty} \operatorname{Tr}(Fe^{-t\Delta_q})$ and so,

$$L(\zeta, \varphi) = \lim_{t \to \infty} \sum_q (-1)^q \operatorname{Tr}(Fe^{-t\Delta_q}). \qquad (10.6)$$

Let us analyze this expression more closely.

PROPOSITION 10.7 *For all values of $t > 0$,*
$$\sum_q (-1)^q \operatorname{Tr}(F e^{-t\Delta_q}) = L(\zeta, \varphi).$$

PROOF It is enough to prove that $\sum_q (-1)^q \operatorname{Tr}(F e^{-t\Delta_q})$ is constant in t. We differentiate, getting
$$\sum_q (-1)^{q+1} \operatorname{Tr}(F(dd^* + d^*d) e^{-t\Delta_q}).$$

Now $dF = Fd$, so $\operatorname{Tr}(Fdd^* e^{-t\Delta_q}) = \operatorname{Tr}(dFd^* e^{-t\Delta_q})$. Assume for the moment that we can apply (8.8) to the operators d and $Fd^* e^{-t\Delta_q}$ (which is not really allowed, since d is unbounded); then
$$\begin{aligned}
\operatorname{Tr}(dF d^* e^{-t\Delta_q}) &= \operatorname{Tr}(F d^* e^{-t\Delta_q} d) \\
&= \operatorname{Tr}(F d^* d e^{-t\Delta_{q-1}}), \quad \text{since } \Delta_q d = d \Delta_{q-1}.
\end{aligned}$$

The terms in the sum for the derivative therefore cancel in pairs, giving 0. (This is "supersymmetry".) It remains to verify that we may apply 8.8, which we do by reducing to the bounded case, as follows:
$$\begin{aligned}
\operatorname{Tr}(dF d^* e^{-t\Delta_q}) &= \operatorname{Tr}(dF d^* e^{-t\Delta_q/2} e^{-t\Delta_q/2}) \\
&= \operatorname{Tr}(e^{-t\Delta_q/2} dF d^* e^{-t\Delta_q/2}) \quad \text{(by 8.8)} \\
&= \operatorname{Tr}(F d^* e^{-t\Delta_q/2} e^{-\Delta_q/2} d) \quad \text{(by 8.8 again)} \\
&= \operatorname{Tr}(F d^* e^{-t\Delta_q} d) \quad \square
\end{aligned}$$

PROPOSITION 10.8 *If φ has no fixed points, then the Lefschetz number $L(\zeta, \varphi)$ is*

PROOF By (10.7)
$$L(\zeta, \varphi) = \sum_q (-1)^q \operatorname{Tr}(F e^{-t\Delta_q}),$$

for any $t > 0$. Look at the behavior of this expression for small t. If $k_t^q(m_1, m_2)$ denotes the heat kernel corresponding to $e^{-t\Delta_q}$, then $F e^{-t\Delta_q}$ is a smoothing operator with kernel
$$(m_1, m_2) \to {}_1\zeta \cdot k_t^q(\varphi(m_1), m_2)$$

where $_1\zeta$ denotes ζ acting on the first variable in the tensor product bundle $S \boxtimes S^*$ of which k is a section. Therefore, by (8.12),

$$\mathrm{Tr}(Fe^{-t\Delta_q}) = \int_M \mathrm{tr}(_1\zeta k_t^q(\varphi(m), m)\,\mathrm{vol}(m).$$

Now the assumption that φ has no fixed points means that its graph $\{(\varphi(m), m) : m \in M\}$ never meets the diagonal in $M \times M$. Therefore, by the asymptotic expansion 7.15, or by the localization result 7.24, $k_t^q(\varphi(m), m)$ tends to 0, uniformly in m, as $t \to 0$. The result follows. □

EXAMPLE 10.9 Any holomorphic automorphism of complex projective space has a fixed point. To prove this, let M be a complex projective space. Then M is a Kähler manifold, so its Dolbeault complex is a Dirac complex. It is known that the Dolbeault cohomology of M is

$$H_{\bar{\partial}}^q(M) = \begin{cases} \mathbb{C} & (q = 0) \\ 0 & (q > 0). \end{cases}$$

A holomorphic automorphism of M induces a geometric endomorphism of the Dolbeault complex, which must act identically on H^0, and so has strictly positive Lefschetz number.

The fixed-point contributions

If there are fixed points, the argument of (10.8) shows that the Lefschetz number is given by a sum of contributions coming from the components of the fixed point set. To get a Lefschetz formula, we must work out these contributions! We will only consider the easiest case, that of "simple fixed points". Let $T_m\varphi$ denote the endomorphism of the tangent space $T_m M$ at a fixed point m induced by φ.

DEFINITION 10.10 The fixed point m is *simple* if $\det(1 - T_m\varphi) \neq 0$.

Another way of saying this is that a simple fixed point is one where the graph of φ cuts the diagonal transversally. It follows that there can be only finitely many such fixed points. To work out the contribution from a simple fixed point, we will use the following lemma:

LEMMA 10.11 Let T be an $n \times n$ matrix. Then for any $t > 0$,

$$\frac{1}{(4\pi t)^{n/2}} \int_{\mathbb{R}^n} e^{-|x-Tx|^2/4t} \, d^n x = \frac{1}{|\det(1-T)|}.$$

PROOF Let $A = (1-T)(1-T^*)$, so $|x - Tx|^2 = (Ax, x)$. Let $\lambda_1, \ldots, \lambda_n$ be the eigenvalues of A; then by an orthogonal change of co-ordinates on $\mathbb{R}^n$

$$\frac{1}{(4\pi t)^{n/2}} \int_{\mathbb{R}^n} e^{-|x-Tx|^2/4t} \, d^n x = \frac{1}{(4\pi t)^{n/2}} \int_{\mathbb{R}^n} \exp\left(\frac{-\lambda_1 x_1^2 - \cdots - \lambda_n x_n^2}{4t}\right) \cdot d^n x$$

$$= \prod_{j=1}^{n} \left(\frac{1}{\sqrt{4\pi t}} \int_{-\infty}^{\infty} e^{-\lambda_j x^2/4t} \, dx\right) = \prod_{j=1}^{n} \frac{1}{|\lambda_j|^{\frac{1}{2}}} = \frac{1}{|\det(1-T)|}. \quad \square$$

THEOREM 10.12 (ATIYAH-BOTT) Let (ζ, φ) be a geometric endomorphism of a Dirac complex (S, d), having only simple fixed points. Then the Lefschetz number of (ζ, φ) is given by the following formula:

$$L(\zeta, \varphi) = \sum_{\varphi(m)=m} \sum_{q=0}^{n} \left(\frac{(-1)^q \operatorname{tr}(\zeta_q(m))}{|\det(1 - T_m \varphi)|}\right).$$

PROOF The argument of (10.8) shows that in order to evaluate the integral

$$\operatorname{Tr}(F e^{-t \Delta_q}) = \int_M \operatorname{tr}({}_1 \zeta_q k_t^q(\varphi(m), m)) \operatorname{vol}(m)$$

asymptotically as $t \to 0$, we need only integrate over arbitrarily small neighborhoods of the fixed points. Therefore, we work in geodesic co-ordinates having their origin at a fixed point. Then we have the following approximations where $T = T_0 \varphi$ and $g = \det(g_{ij})$:

$$\zeta_q(x) = \zeta_q(0) + O(|x|)$$
$$\varphi(x) = Tx + O(|x|^2)$$
$$g(x) = 1 + O(|x|).$$

Moreover, by truncating the asymptotic expansion (7.15) at a point where the error of the expansion is uniformly of order t, we obtain

$$k_t^q(m', m) = \frac{1}{(4\pi t)^{n/2}} \exp\left(\frac{-d(m', m)^2}{4t}\right) \cdot (\Theta_0(m', m) + O(t)) + O(t)$$

where $\Theta_0(m, m) = 1$.

Since $\det(1 - T) \neq 0$, there is a constant $\delta > 0$ such that
$$|x - Tx|^2 \geq \delta |x|^2.$$
Moreover, $d(\varphi(x), x)^2 = |x - Tx|^2 + O(|x|^3)$ and $\Theta_0(\varphi(x), x) = 1 + O(|x|)$, so the asymptotic expansion formula gives
$$k_t^q(\varphi(x), x) = \frac{1}{(4\pi t)^{n/2}} e^{-|x-Tx|^2/4t} \left(1 + O(|x|) + O(t) + O(|x|^3/t)\right) + O(t).$$
Therefore
$$\left| {}_1\zeta_q(x) k_t^q(\varphi(x), x) \sqrt{g(x)} - \frac{\zeta_q(0)}{(4\pi t)^{n/2}} e^{-|x-Tx|^2/4t} \right| \tag{10.13}$$
$$\leq \frac{1}{(4\pi t)^{n/2}} e^{-\delta|x|^2/4t} \left(O(|x|) + O(t) + O\left(\frac{|x|^3}{t}\right) \right) + O(t).$$
By integration one finds that the L^1 norm of
$$\frac{1}{(4\pi t)^{n/2}} e^{-\delta|x|^2/4t} \cdot |x|^a t^b$$
is of order $t^{a/2+b}$; so the right-hand side of (10.13) is of order $t^{\frac{1}{2}}$ in L^1 norm as $t \to 0$. Therefore, as $t \to 0$,
$$\int \operatorname{tr}\left({}_1\zeta_q(x) k_t^q(\varphi(x), x)\right) \sqrt{g(x)} \, d^n x \to \int_{\mathbf{R}^n} \operatorname{tr}\left(\zeta_q(0) \cdot \frac{1}{(4\pi t)^{n/2}} e^{-|x-Tx|^2/4t}\right) d^n x$$
and by Lemma 10.11, the right hand side is equal to $\frac{\operatorname{tr}(\zeta_q(0))}{|\det(1-T)|}$. So
$$\operatorname{Tr}(F e^{-t\Delta_q}) \to \sum_{\varphi(m)=m} \frac{\operatorname{tr}(\zeta_q(m))}{|\det(1 - T_m \varphi)|}.$$
Now the result follows by (10.7). □

Many applications of this result may be found in the paper of Atiyah and Bott [2]. Here are a couple of examples.

EXAMPLE 10.14 Suppose that our Dirac complex is in fact the de Rham complex. Then the maps $\zeta_q \colon \varphi^* S_q \to S_q$ are just the exterior powers $\bigwedge^q T^*$ of the dual of the tangent map $T = T\varphi$ to φ. Now in general for any linear transformation T,
$$\sum_q (-1)^q \operatorname{tr}(\bigwedge^q T) = \det(1 - T);$$

as one can check easily by recalling that $\operatorname{tr}(\Lambda^q T)$ is the q'th elementary symmetric function of the eigenvalues of T. Therefore the contribution from the fixed point m in the Atiyah-Bott formula is just

$$\frac{\det(1 - T_m^*\varphi)}{|\det(1 - T_m\varphi)|} = \operatorname{sgn}\det(1 - T_m\varphi).$$

We therefore recover the original Lefschetz theorem:

$$L(\varphi) = \sum_{\varphi(m)=m} \operatorname{sgn}\det(1 - T_m\varphi).$$

EXAMPLE 10.15 Now let φ be a holomorphic automorphism of a compact Kähler manifold M. As explained in Example 10.9, φ induces a geometric endomorphism of the Dolbeault complex of M, whose Lefschetz number is defined to be the *holomorphic Lefschetz number* of φ, denoted $L_{\bar{\partial}}\varphi$.

To work out the local contribution in the Atiyah-Bott formula, we must do a little linear algebra. The tangent space at a fixed point is a real vector space V of even dimension equipped with a complex structure J (multiplication by $i = \sqrt{-1}$), and the tangent map $T : V \to V$ is *complex* linear, that is $JT = TJ$. Then $V \otimes_{\mathbf{R}} \mathbf{C}$ decomposes as a sum $P \oplus Q$ of eigenspaces for J of eigenvalues $\pm i$, and correspondingly $T \otimes_{\mathbf{R}} 1$ decomposes as $T \oplus \overline{T}$; P is isomorphic as a complex vector space to V and Q is isomorphic to $\overline{V}$. Now the determinant appearing in the fixed point formula is $\det_{\mathbf{R}}(1 - T)$, the determinant of $1 - T$ considered as a *real* linear map. But

$$\det_{\mathbf{R}}(1 - T) = \det_{\mathbf{C}}((1 - T) \otimes 1) = \det_{\mathbf{C}}(1 - T)\det_{\mathbf{C}}(1 - \overline{T})$$

by the decomposition. On the other hand, $\zeta_q = \Lambda^q \overline{T}^*$, acting on the bundle $\Lambda^q Q^*$, so

$$\sum (-1)^q \operatorname{tr}(\zeta_q) = \det_{\mathbf{C}}(1 - \overline{T}).$$

Thus we get Atiyah and Bott's *holomorphic Lefschetz theorem*:

$$L_{\bar{\partial}}(\varphi) = \sum_{\varphi(m)=m} \frac{1}{\det_{\mathbf{C}}(1 - T_m\varphi)}$$

in the case of simple fixed points.

We required M to be a Kähler manifold because only for such manifolds is the Dolbeault complex a Dirac complex. However, by working with generalized Dirac operators one can prove the holomorphic Lefschetz theorem for all compact complex manifolds.

Notes

The Lefschetz theorem and many applications may be found in Atiyah and Bott [2]. For the "heat equation" proof, see Dieudonné [25, Volume IX], or Gilkey [34].

Exercises

QUESTION 10.16 Let M be a compact oriented n-manifold, and let X be a vector field on it. A point $m \in M$ such that $X(m) = 0$ is called a *critical point* of X, and it is *non-degenerate* if the derivative of X at m (which is an $n \times n$ matrix in local co-ordinates) is non-singular. The sign (± 1) of the determinant of this matrix is the *index* of the critical point.

Prove Hopf's theorem, that the sum of the indices of the critical points of a vector field having only non-degenerate critical points is equal to the Euler characteristic of M. (Apply the Lefschetz theorem to the flow generated by X.)

QUESTION 10.17 (ATIYAH-BOTT) Consider the endomorphism φ of $\mathbb{CP}^n$ defined in terms of homogeneous co-ordinates $(z_0, \ldots, z_n)$ by

$$(z_0, \ldots, z_n) = (\gamma_0 z_0, \ldots, \gamma_n z_n)$$

where the γ's are distinct non-zero complex numbers. From the holomorphic Lefschetz theorem for φ deduce the Legendre interpolation formula

$$1 = \sum_{i=0}^{n} \frac{\gamma_i^n}{\prod_{j \neq i}(\gamma_i - \gamma_j)}.$$

CHAPTER 11

The index problem

Gradings and Clifford bundles

Recall that a module W over a Clifford algebra $\mathrm{Cl}(V)$ is said to be *graded* if it is provided with a decomposition $W = W_+ \oplus W_-$ such that Clifford multiplication by any $v \in V$ interchanges the summands W_+ and W_-. A Clifford bundle S on a Riemannian manifold is *graded* if it is provided with a decomposition $S = S_+ \oplus S_-$ which respects the metric and connection and makes each fiber S_x a graded Clifford module over $\mathrm{Cl}(T_xM)$. It is equivalent to say that S is provided with an involution ε (the *grading operator*) which is self-adjoint, parallel[1] and such that $\varepsilon c(v) + c(v)\varepsilon = 0$ for every tangent vector v. The sub-bundles $S_\pm$ are the ± 1 eigenspaces of ε.

If S is a graded Clifford bundle, then the algebra of bounded operators on $L^2(S)$ is a superalgebra in the sense of definition 4.1.

DEFINITION 11.1 Let A be a trace-class operator on $L^2(S)$, where S is a graded Clifford bundle. Then the *supertrace* of A is defined to be

$$\mathrm{Tr}_s(A) = \mathrm{Tr}(\varepsilon a)$$

where ε is the grading operator.

It is easy to check, by reduction to the corresponding property of the ordinary trace (8.8), that the supertrace vanishes on supercommutators $[A, B]_s$ provided that one of A and B is trace-class. There is an obvious analogue of 8.12, namely:

PROPOSITION 11.2 *Let A be a smoothing operator on $L^2(S)$ with kernel $k \in C^\infty(S \boxtimes S^*)$; then*

$$\mathrm{Tr}_s(A) = \int_M \mathrm{tr}_s(k(x,x))\,\mathrm{vol}(x)$$

where the 'local supertrace' $\mathrm{tr}_s(a)$, $a \in \mathrm{End}(S_x)$, *is defined to be* $\mathrm{tr}(\varepsilon a)$.

---

[1] That is, it commutes with covariant differentiation.

We will be concerned with Clifford bundles over *even-dimensional, oriented* manifolds, say of dimension $2m$. For such a bundle there is always a canonical grading; the volume element ω in the Clifford algebra (4.4) has $\omega^2 = (-1)^m$ and anticommutes with all the Clifford generators, so that the Clifford action of $i^m \omega$ defines a grading operator ε_0 on any Clifford bundle. However, other gradings are possible. If ε is another such grading then $\varepsilon\varepsilon_0$ is a self-adjoint involution which commutes with the whole Clifford algebra; thus we obtain

LEMMA 11.3 *Any graded Clifford bundle S is split into a direct sum of two graded Clifford sub-bundles, on one of which $\varepsilon = \varepsilon_0$ and on the other of which $\varepsilon = -\varepsilon_0$. (We refer to these as the* canonically graded *and* anticanonically graded *parts of S respectively.)*

Because of the existence of this direct sum decomposition, it is often sufficient to restrict our attention to canonically graded Clifford bundles.

We want to analyze the local supertrace $\operatorname{tr}_s(a)$, $a \in \operatorname{End}(S_x)$, which appears in 11.2 above, using the representation theory of the Clifford algebra. The uniqueness of the spin representation provides a decomposition (4.12) $S_x = \Delta \otimes V$ where Δ denotes the spin-representation and V is an auxiliary vector space, and there is a corresponding decomposition on the endomorphism level

$$\operatorname{End}(S_x) = \operatorname{Cl}(T_x M) \otimes \operatorname{End}(V), \qquad \operatorname{End}(V) = \operatorname{End}_{\operatorname{Cl}}(S_x).$$

To say that S_x is canonically graded is simply to say that the grading of S_x is given as $(\Delta_+ \otimes V) \oplus (\Delta_- \otimes V)$, where $\Delta_\pm$ are the positive and negative half-spin representations. Let $\tau_s \colon \operatorname{Cl} \to \mathbb{C}$ denote the supertrace of the action of the Clifford algebra on the spin representation; then the above discussion proves

PROPOSITION 11.4 *Let $a = c \otimes F$ be an endomorphism of S_x, where $c \in \operatorname{Cl}(T_x M)$ and $F \in \operatorname{End}_{\operatorname{Cl}}(S_x)$. Suppose that S is canonically graded. Then*

$$\operatorname{tr}_s(a) = \tau_s(c) \operatorname{tr}^{S/\Delta}(F)$$

where the relative trace $\operatorname{tr}^{S/\Delta}$ is defined in 4.13.

In order to apply this proposition effectively one needs to know how to compute τ_s. The following lemma does this. To state it, let $\{e_1, \ldots, e_{2m}\}$ be an orthonormal basis of $\mathbb{R}^{2m}$. If $E \subset \{1, \ldots, 2m\}$, then $\tilde{E}$ will denote the element $\prod_{i \in E} e_i$ of $\mathrm{Cl}(\mathbb{R}^{2m})$. Recall that the $\tilde{E}$ form a linear basis for the Clifford algebra $\mathrm{Cl}(\mathbb{R}^{2m})$.

LEMMA 11.5 Let $c = \sum_E c_E \tilde{E}$ be an element of $\mathrm{Cl}(\mathbb{R}^{2m})$. Then the super-trace $\tau_s(c)$ of c, considered as an endomorphism of the spin representation, is equal to $(-2i)^m c_{12\ldots 2m}$. That is, the supertrace is equal (up to a scalar multiple) to the 'top degree part' of c.

PROOF By definition of the super-trace $\tau_s(c) = \tau(i^m \omega c)$, where τ denotes the ordinary trace on the spin representation. It is therefore enough to prove that

$$\tau(c) = 2^m c_\emptyset .$$

To do this, we must prove that

$$\tau(\tilde{E}) = \begin{cases} 2^m & \text{if } E = \emptyset \\ 0 & \text{otherwise.} \end{cases}$$

Clearly $\tilde{\emptyset} = 1$ acts on Δ with trace $\dim(\Delta) = 2^m$. Now if $E \neq \emptyset$, consider $\tilde{E}$ acting on the Clifford algebra itself by left multiplication. As a representation, the Clifford algebra is equal to $\Delta \otimes \Delta^*$, with left action on the first factor Δ. So the trace of $\tilde{E}$ on Δ is equal to 2^{-m} times the trace of $\tilde{E}$ on the Clifford algebra itself. But $\tilde{E}$ permutes the basis elements without fixed point, so $\mathrm{tr}(\tilde{E}) = 0$. □

REMARK 11.6 One can also deduce this lemma from the computation of the character table for the finite group E_{2m}, which we carried out in Chapter 4.

Graded Dirac operators

The Dirac operator of a graded Clifford bundle anticommutes with the grading operator, and so maps sections of $S_\pm$ to sections of $S_\mp$. We may therefore think of the Dirac operator as coming from a Dirac complex of length 2

$$C^\infty(S_+) \xrightarrow{D_+} C^\infty(S_-)$$

where D_+ is the restriction of D to sections of S_+ and its adjoint $D_+^* = D_-$ is the restriction of D to sections of S_-. The Euler characteristic of this complex is an important invariant called the *index* of D. More formally

DEFINITION 11.7 The *index* of a graded Dirac operator D is the difference
$$\mathrm{Ind}(D) = \dim \ker D_+ - \dim \ker D_-.$$

EXAMPLE 11.8 As a simple example, consider the de Rham operator $D = d + d^*$ with the grading operator defined by $\varepsilon = (-1)^q$ on $\Omega^q(M)$. Then by Hodge theory, the index of D is simply the Euler characteristic of M in the sense of topology. Notice that this grading, which we will call the *Euler grading* of the de Rham operator, is neither canonical nor anticanonical.

It is immediate from the definitions that we have $\mathrm{Ind}(D) = \mathrm{Tr}_s(P)$, where P denotes the orthogonal projection onto $\ker(D)$. More generally we have

PROPOSITION 11.9 Let f be any rapidly decreasing smooth function on $\mathbb{R}^+$ with $f(0) = 1$. Then $\mathrm{Ind}(D) = \mathrm{Tr}_s(f(D^2))$.

PROOF Since the spectrum of D is discrete, the projection P onto the kernel of D can be written as $f(D)$ for some appropriately chosen compactly supported smooth function f with $f(0) = 1$. It therefore suffices to prove that $\mathrm{Tr}_s(g(D^2)) = 0$ when g is smooth and rapidly decreasing and $g(0) = 0$. We may write $g(x) = xh(x)$ for some rapidly decreasing function h; and we may further write $h(x) = h_1(x)h_2(x)$ where both h_1 and h_2 are rapidly decreasing. But now
$$g(D^2) = D^2 h(D^2) = \tfrac{1}{2}[Dh_1(D), Dh_2(D)]_s$$
and so $\mathrm{Tr}_s\, g(D^2) = 0$ since the supertrace vanishes on supercommutators. □

REMARK 11.10 Because of its historical importance we give a variation of this proof. For an eigenvalue λ of D^2, let $n_+(\lambda)$ denote the dimension of the λ-eigenspace of the restriction of D^2 to S_+, and similarly for $n_-(\lambda)$. Then clearly
$$\mathrm{Tr}_s\, f(D^2) = \sum_\lambda f(\lambda)(n_+(\lambda) - n_-(\lambda)) = \mathrm{Ind}(D) + \sum_{\lambda > 0} f(\lambda)(n_+(\lambda) - n_-(\lambda)).$$
But for $\lambda \neq 0$, the operator D gives an isomorphism between the λ-eigenspaces of D^2 on S_+ and on S_-, so $n_+(\lambda) = n_-(\lambda)$, giving the result.

The special case
$$\mathrm{Ind}(D) = \mathrm{Tr}_s\, e^{-tD^2} \tag{11.11}$$

relating the index to the heat equation is known as the *McKean-Singer* formula [51].

REMARK 11.12 Proposition 11.9 applies to rapidly decreasing functions f. But a study of its proof reveals that it is enough that $f(x) = O(x^{-N})$, where N is some constant, depending on the dimension, which is sufficiently large that the operators $Dh_1(D)$ and $Dh_2(D)$ appearing in the proof can be chosen to have continuous kernels and so to be Hilbert-Schmidt (see 5.32 and 8.9). This will be important in a moment.

We study the variation of the index as the operator D varies. Let D_t, $t \in [0,1]$, be a continuous family of graded Dirac operators on (M, S); by this we mean that the Riemannian metric, the Clifford action, and the metric and connection on S are all varying continuously with t (in such a way as to preserve the compatibility condition j. Then $t \mapsto D_t$ is a continuous map from $[0,1]$ to $B(W^{k+1}, W^k)$ for any k, and the operators D_t all satisfy the elliptic estimates

$$\|s\|_{k+1}^2 \leq C_k(\|s\|_k^2 + \|Ds\|_k^2)$$

with a constant C_k which is uniform in t.

PROPOSITION 11.13 *Let D_t be a continuous family of graded Dirac operators, as above; then $\mathrm{Ind}(D_0) = \mathrm{Ind}(D_1)$.*

PROOF The resolvents $(D_t \pm i)^{-1}$ map W^k to W^{k+1} for any k, as is shown by elliptic estimates. Moreover the maps $t \mapsto (D_t \pm i)^{-1}$ are continuous from $[0,1]$ to $B(W^k, W^{k+1})$ for any $k \geq 0$. To see this we use the resolvent formula

$$(D_t + i)^{-1} - (D_{t'} + i)^{-1} = (D_t + i)^{-1}(D_{t'} - D_t)(D_{t'} + i)^{-1}$$

and the uniformity in the elliptic estimates which shows that the $B(W^k, W^{k+1})$ norm of $(D_t + i)^{-1}$ is bounded independent of t. It follows that $(1 + D_t^2)^{-N}$ is continuous from $[0,1]$ to $B(W^k, W^{k+2N})$. When N is large enough the inclusion $W^{k+2N} \to W^k$ is a Hilbert-Schmidt operator (exercise 8.22). Taking such an N we deduce that $(1 + D_t^2)^{-2N}$ is a trace class operator and that its trace, or its supertrace, vary continuously with t. Thus by 11.9 and the subsequent remarks,

$$\mathrm{Ind}(D) = \mathrm{Tr}_s((1 + D_t^2)^{-2N})$$

varies continuously with t. Since, however, the index is an integer, it must be constant. □

This fundamental stability property shows that the index is a *topological invariant*: it depends only on homotopy-theoretic data about the manifold M and bundle S. The *index problem* which was solved by Atiyah and Singer in the early 1960's was this: compute the index in terms of the conventional invariants of algebraic topology, namely characteristic classes associated to the bundle S and to the tangent bundle of M. The index theorem made possible a vigorous commerce between analysis and topology. On the one hand, information about the index derived from PDE theory — sometimes even the rather minimal information that $\text{Ind}(D)$ is an integer — could be used to constrain the characteristic classes and hence the topology of M. On the other hand, topological conditions could force the existence of solutions to differential equations — holomorphic functions, for example — which solutions might then be used in further geometric constructions. Milnor's construction of the exotic spheres [53] and the Kodaira embedding theorem (see [36]) are examples of these two phenomena which predate the general form of the index theorem itself.

The original solutions to the index problem depended on the use of algebraic topology (either cobordism theory or K-theory) to organize the possible pairs (M, S) into some kind of group, and thus to reduce the proof of the index theorem to a check on some specific generators. Thus they were essentially global and topological in nature. In this book we will prove the index theorem by an alternative method, which is based on the McKean-Singer formula [51]. McKean and Singer pointed out that we have an asymptotic expansion for the heat kernel which is in principle locally computable, and that therefore formula 11.11 can be used to relate the index to the local super-trace of certain coefficients appearing in that expansion. The relevant coefficients are, however, almost impossible to compute by brute force. Following an idea of Gilkey, namely to use Invariant Theory to study the polynomials in the curvature which might possibly arise in the expansion, a proof of the index theorem using the asymptotic expansion was given by Atiyah, Bott and Patodi in [3]. But Gilkey's idea suffered from the defect that Invariant Theory could only determine the coefficients up to a (finite) number of arbitrary constants; it was still necessary to compute examples in order to fix the values of the constants.

The situation was transformed by the appearance of Getzler's paper [31]. Getzler showed that the computations needed in the asymptotic-expansion method can be rendered quite tractable by paying careful attention to the rôle of the Clifford algebra. In his method, the fact 11.5 that the local supertrace corresponds to the 'top degree' part in the Clifford algebra is of crucial importance. It is used to reduce the computations to those in a simple local model, essentially the harmonic oscillator; and the function $a/\sinh(2at)$ appearing in Mehler's formula 9.12 for the harmonic oscillator heat kernel is revealed to lead to the appearance of the $\hat{A}$-genus (2.28), the Pontrjagin genus associated to the function $w/\sinh w$ where $w = \frac{1}{2}\sqrt{z}$, on the topological side of the index theorem for the Dirac operator. The rest of this chapter and the next give an exposition of Getzler's proof.

The heat equation and the index theorem

With notation as above, recall the asymptotic expansion 7.15 for the heat kernel $_t$ associated to the smoothing operator e^{-tD^2}. Using 11.2 this gives us

$$\mathrm{Tr}_s(e^{-tD^2}) \sim \frac{1}{(4\pi t)^{n/2}} \left(\int \mathrm{tr}_s \Theta_0 \,\mathrm{vol} + t \int \mathrm{tr}_s \Theta_1 \,\mathrm{vol} + \ldots \right).$$

But $\mathrm{Tr}_s(e^{-tD^2})$ is in fact constant, and equal to the index of D, by the McKean-Singer formula. So we get:

PROPOSITION 11.14 *The index of the graded Dirac operator D is zero if $n(= \dim M)$ is odd, and is equal to*

$$\frac{1}{(4\pi)^{n/2}} \int \mathrm{tr}_s \Theta_{n/2} \,\mathrm{vol}$$

if n is even, where the asymptotic expansion coefficient $\Theta_{n/2}$ is a certain algebraic expression in the metrics and connection coefficients and their derivatives.

Here is a non-trivial corollary.

COROLLARY 11.15 *The index is multiplicative under coverings; i.e. if $\tilde{M}$ is a k-fold covering of M, and $\tilde{S}$, $\tilde{D}$ are the natural lifts of S and D to $\tilde{M}$, then $\mathrm{Ind}(\tilde{D}) = k\,\mathrm{Ind}(D)$.*

This is immediate from 11.14, since $\Theta_{n/2}$ is a local expression which is the same on M as on $\tilde{M}$. It is not obvious from the definition of the index.

EXAMPLE 11.16 Let us consider the case of a 2-dimensional Riemannian manifold M, and let $D = d + d^*$ be the de Rham operator equipped with the Euler grading 11.8 by the degree (mod 2) of forms. Then by the above, the index of D is equal to

$$\frac{1}{4\pi} \int \left(\mathrm{tr}\, \Theta_1^0 - \mathrm{tr}\, \Theta_1^1 + \mathrm{tr}\, \Theta_1^2 \right) \mathrm{vol}$$

where the superscript on Θ_1 denotes the degree of differential forms. Now we use our calculation in 7.19, which gives

$$\Theta_1^i = \tfrac{1}{6}\kappa \cdot 1 - \mathsf{K}^i$$

where K is the Clifford-contracted curvature operator appearing in the Weitzenbock formula. Now $\mathsf{K}^0 = 0$ and so $\mathsf{K}^2 = 0$ also by Hodge duality; on the other hand, K^1 is the Ricci curvature operator by 6.8. Thus we get

$$\mathrm{tr}\, \Theta_1^0 = \mathrm{tr}\, \Theta_1^2 = \tfrac{1}{6}\kappa, \quad \mathrm{tr}\, \Theta_1^1 = \tfrac{1}{6}\kappa \cdot 2 - \kappa = -\tfrac{2}{3}\kappa,$$

and so we finally obtain

$$\mathrm{Ind}(D) = \frac{1}{4\pi} \int \kappa\, \mathrm{vol}$$

and bearing in mind that $\mathrm{Ind}(D)$ is the Euler characteristic and that κ is twice the Gaussian curvature, we recognize the Gauss-Bonnet theorem.

Notes

References for the original proofs of the index theorem are the seminar by Palais [57] and of course the series of papers by Atiyah, Singer, and Segal [9, 8, 10] in the *Annals of Mathematics*. Atiyah's Collected Works were published since the first edition of this book and are a must-read for anyone interested in acquiring a deeper understanding of the index theorem and the ideas surrounding it.

Seeley [67] gives a historical account of the development of index theory from the point of view of singular integral operators.

Exercises

QUESTION 11.17 Let D be a graded Dirac operator, and consider the associated Dirac complex of length two.

(i) Show that the index of D is simply the Lefschetz number of the identity map on this complex.

(ii) Relate the McKean-Singer formula to the Lefschetz number formula of 10.7.

(iii) Since the index is a Lefschetz number, why can't we apply the Atiyah-Bott Lefschetz theorem of the last chapter to calculate it?

QUESTION 11.18 Let H_1 and H_2 be Hilbert spaces and let $A: H_1 \to H_2$ be a bounded operator. Suppose that there is a "parametrix" for A, that is an operator $Q: H_2 \to H_1$ such that the operators $AQ - 1$ and $QA - 1$ are compact. Show that $\ker A$ and $\ker A^*$ are finite-dimensional, and that the 'Fredholm index' $\mathrm{Ind}(A) = \dim \ker A - \dim \ker A^*$ is a locally constant function of A.

QUESTION 11.19 Now let $A: H_1 \to H_2$ have a parametrix Q such that, for some positive integer p, $(AQ-1)^p$ and $(QA-1)^p$ are trace-class operators. Show that

$$\mathrm{Ind}\, A = \mathrm{Tr}((QA-1)^p) - \mathrm{Tr}((AQ-1)^p).$$

QUESTION 11.20 Let D be a Dirac operator on a compact n-manifold, and let k_t be the corresponding heat kernel. Consider the differential n-form α_t defined by $\alpha_t(m) = \mathrm{tr}_s(k_t(m,m))\,\mathrm{vol}(m)$, where tr_s denotes the *local* super-trace. Prove that the derivative $\partial \alpha_t / \partial t$ is an exact n-form, and thereby obtain yet another proof of the McKean-Singer formula.

QUESTION 11.21 Use the heat equation method to prove the Riemann-Roch formula for a Riemann surface (same method as 11.16 above). This is due to Kotake [45].

QUESTION 11.22 Let V be a complex vector bundle over a compact manifold M. Show that there is a smooth function $e: M \to M_N(\mathbb{C})$ for some N, which is a self-adjoint projection ($e^2 = e$) and is such that the range of e is a sub-bundle of $M \times \mathbb{C}^N$ isomorphic to V.

149

Show further that the equation
$$\nabla v = e(dv)$$
defines a connection on the range of e.

QUESTION 11.23 Let S be a Clifford bundle over the compact manifold M, and let D be the corresponding Dirac operator. Let V be a vector bundle over M, represented as the range of a projection-valued function e, and equipped with its natural connection (see question 11.22). Show that the Dirac operator D_V on S with coefficients in V (3.24) is given by $D_V = e(D \otimes 1)e$.

Now suppose further that $\mathrm{Ker}(D) = 0$. Show that D^{-1} is a bounded operator on $L^2(S)$. Show also that $Q = e(D^{-1} \otimes 1)e$ is a parametrix for D_V, and that in fact for sufficiently large p,
$$\mathrm{Ind}(D_V) = \mathrm{Tr}_s(QD_V - 1)^p).$$

QUESTION 11.24 (CONNES [21, 22]). In the situation of question 11.23, the following expression is called the (cyclic) *character* of D:
$$\tau(f_0, f_1, \ldots, f_{2p}) = \mathrm{Tr}_s(D^{-1}[D, f_0]D^{-1}[D, f_1]\ldots D^{-1}[D, f_{2p}])$$
where $f_0, \ldots, f_{2p}$ are (possibly matrix-valued) smooth functions on M.
(a) Show that the expression makes sense (i.e. that the operator in the brackets is of trace-class) if p is large enough.
(b) Show that τ is a *cocycle*: i.e. that for any functions $f_0, \ldots, f_{2p+1}$,
$$\sum_{j=0}^{2p}(-1)^j\tau(f_0, f_1, \ldots, f_jf_{j+1}, \ldots, f_{2p+1}) = \tau(f_{2p+1}f_0, f_1, \ldots, f_{2p}).$$
(c) Show that if e is a projection-valued function corresponding to a vector bundle V, then $\mathrm{Ind}(D_V) = \tau(e, e, \ldots, e)$.

(This computation lies at the beginning of the relationship between index theory and Connes' *cyclic cohomology*.)

CHAPTER 12

The Getzler calculus and the local index theorem

This chapter gives the heart of the proof of the Index Theorem. We will study the so-called *symbolic calculus* for operators on bundles of Clifford modules. The idea is to provide a systematic way of investigating the 'top order part' of an operator or a family of operators. For instance, our proof of the Weyl asymptotic formula 8.16 was based on our knowledge that the 'top order part' of the heat kernel on a manifold is simply the heat kernel on Euclidean space. Getzler's innovation was the introduction of a sophisticated notion of 'order', with respect to which the index form — discussed at the end of the last chapter — naturally appears as a 'top order part'.

Filtered algebras and symbols

In this section we will be concerned with algebras over $\mathbb{C}$. Recall that an *algebra* A is a vector space equipped with a bilinear, associative product. A *graded algebra* is an algebra provided with a direct sum decomposition $A = \bigoplus A^m$, such that $A^m \cdot A^{m'} \subseteq A^{m+m'}$. Familiar examples of graded algebras include the exterior algebra $\Lambda^* V$ over a vector space, and the polynomial algebra $\mathbb{C}[t]$.

The notion of graded algebra should be contrasted with that of *filtered algebra*:

DEFINITION 12.1 A *filtration* of an algebra A is a family of subspaces A_m, $m \in \mathbb{Z}$, with $A_m \subseteq A_{m+1}$, and such that $A_m \cdot A_{m'} \subseteq A_{m+m'}$, for all $m, m' \in \mathbb{Z}$. An algebra provided with a filtration is called a *filtered algebra*.

EXAMPLE 12.2 The algebra $\mathfrak{D}(M)$ of differential operators acting on functions on a manifold M is a filtered algebra, with $\mathfrak{D}_m(M)$ equal to the space of differential operators of order $\leqslant m$.

EXAMPLE 12.3 The Clifford algebra $\mathrm{Cl}(V)$ of an inner product space V is a filtered algebra, with $\mathrm{Cl}_m(V)$ equal to the linear span of the products of m or fewer elements of V.

REMARK 12.4 Clearly, any graded algebra can be regarded as a filtered algebra (just define $A_m = A^0 \oplus \cdots \oplus A^m$). Moreover, any homomorphic image of a filtered algebra is a filtered algebra. These ideas can be put together to generate a filtration on any algebra from an assignment of degrees to members of some generating set. For an example sufficient for our purposes, suppose that A is generated by $B \cup V$, where B is a subalgebra of A, V is a vector subspace, and we want to think of elements of B as having degree zero and elements of V as having degree one. Then there is a surjective homomorphism of algebras

$$\bigotimes_B V = B \oplus (B \otimes V \otimes B) \oplus (B \otimes V \otimes B \otimes V \otimes B) \oplus \cdots \to A$$

and so A inherits a filtration from the tensor algebra. For instance, the Clifford algebra $Cl(V)$ is generated in this way by $B \cup V$ (where $B = \mathbb{C}$), and the filtration defined by these generators is the one in example 12.3 above.

DEFINITION 12.5 Let A be a filtered algebra and let G be a graded algebra. A *symbol map* $\sigma_\bullet : A \to G$ is a family of linear maps $\sigma_m : A_m \to G^m$, such that

(i) If $a \in A_{m-1}$, then $\sigma_m(a) = 0$.
(ii) If $a \in A_m$ and $a' \in A_{m'}$, then $\sigma_m(a)\sigma_{m'}(a') = \sigma_{m+m'}(aa')$.

We refer to (ii) as the *homomorphism-like property* of the symbol.

DEFINITION 12.6 Let A be a filtered algebra. The *associated graded algebra* $G(A)$ is the direct sum

$$G(A) = \bigoplus_m A_m/A_{m-1}$$

with the product operation induced from A (which the reader should verify is well defined).

The quotient maps $A_m \to A_m/A_{m-1}$ give rise to a symbol map $\sigma_\bullet : A \to G(A)$. In fact this is obviously the 'universal' symbol map on A in an appropriate sense, but we will not spell this out in detail.

EXAMPLE 12.7 Let $A = Cl(V)$ be the Clifford algebra of V. Then the associated graded algebra $G(A)$ is the exterior algebra $\wedge^* V$; and the symbol maps $\sigma_\bullet : Cl(V) \to$

$\Lambda^*(V)$ give the top degree part of the linear isomorphism between the Clifford and exterior algebras discussed in 3.23 and exercise 3.32.

EXAMPLE 12.8 Let $A = \mathfrak{D}(M)$, the algebra of differential operators on M. To describe symbols on A, we need to investigate differential operators modulo lower order operators.

Let V be a finite-dimensional vector space, and let $\mathfrak{C}(V)$ denote the algebra of *constant coefficient* differential operators acting on functions on V. Then $\mathfrak{C}(V)$ is a *graded* algebra, its degree m part being made up of homogeneous differential operators of order m. We may form the bundle $\mathfrak{C}(TM)$ whose fiber at a point p is $\mathfrak{C}(T_pM)$; and the space of smooth sections $C^\infty(\mathfrak{C}(TM))$ forms a graded algebra. We will construct a symbol map

$$\sigma_\bullet \colon \mathfrak{D}(M) \to C^\infty(\mathfrak{C}(TM)).$$

Fix $p \in M$. Given a differential operator $T \in A_m$, choose local coordinates x^i with origin p and write

$$T = \sum_{|\alpha| \leqslant m} c_\alpha(x) \frac{\partial^\alpha}{\partial x^\alpha}$$

in terms of these local coordinates. Let $\sigma_{m,p}(T)$ be the constant coefficient differential operator on T_pM obtained by 'freezing coefficients'

$$\sigma_{m,p}(T) = \sum_{|\alpha|=m} c_\alpha(0) \frac{\partial}{\partial x^\alpha}.$$

It is straightforward to check that this definition is independent of the choice of local coordinate system, and it obviously vanishes on operators of order $< m$. Moreover, if $T \in A_m$ and $T' \in A_{m'}$, then $\sigma_{m+m'}(TT') = \sigma_m(T)\sigma_{m'}(T')$, because the commutator of T and multiplication by a smooth function is an operator of order $< m$. The maps $\sigma_{m,p}$ as p varies fit together to give a linear map

$$\sigma_\bullet \colon \mathfrak{D}_m(M) \to C^\infty(\mathfrak{C}^m(TM))$$

which is the desired symbol.

REMARK 12.9   The algebra of differential operators $\mathfrak{D}(M)$ is generated by $B = C^\infty(M)$ in degree zero and $V = \mathcal{X}(M)$ (the vector fields on M, acting by Lie derivative) in degree one; and the filtration on $\mathfrak{D}(M)$ is that determined by these generators. To specify the symbol map completely it is therefore enough to specify its action on the generators. It is easy to see that the symbol $\sigma_0(f)$ of a smooth function f is f itself (thought of as the operator of multiplication by the constant $f(p)$ on T_pM for each p) and similarly the symbol $\sigma_1(X)$ of a vector field X is X itself (thought of as the constant-coefficient first order operator $\partial_{X(p)}$ on T_pM).

Getzler symbols

Now let M be an even-dimensional Riemannian manifold and S a Clifford bundle over it. Recall from 4.12 that we have an isomorphism

$$\operatorname{End}(S) = \operatorname{Cl}(TM) \otimes \operatorname{End}_{\operatorname{Cl}}(S)$$

and we use this isomorphism to make $\operatorname{End}(S)$ into a bundle of filtered algebras, using the standard filtration of $\operatorname{Cl}(TM)$ and giving degree zero to elements of $\operatorname{End}_{\operatorname{Cl}}(S)$. We will call this the *Clifford filtration* on $\operatorname{End}(S)$.

We want to study the algebra $\mathfrak{D}(S)$ of differential operators acting on sections of S. This algebra is generated by Clifford multiplications, covariant derivatives, and sections of the bundle $\operatorname{End}_{\operatorname{Cl}}(S)$.

DEFINITION 12.10   The *Getzler filtration* on $\mathfrak{D}(S)$ is that determined (using the construction of 12.4) by the following assignment of degrees to the generators of $\mathfrak{D}(S)$:

(i) A Clifford module endomorphism of S has degree zero;
(ii) Clifford multiplication $c(X)$, for $X \in \mathcal{X}(M)$, has degree one;
(iii) Covariant differentiation ∇_X, for $X \in \mathcal{X}(M)$, also has degree one.

Whenever we think of $\mathfrak{D}(S)$ as a filtered algebra, we will use this filtration.

We will define a symbol map on $\mathfrak{D}(S)$. Like that of the previous section, its range will be the sections of a certain bundle of differential operators on TM. But the operators no longer have constant coefficients.

DEFINITION 12.11 Let V be a vector space. The notation $\mathfrak{P}(V)$ will denote the algebra of polynomial coefficient differential operators acting on functions on V.

Notice that $\mathfrak{P}(V)$ is a graded algebra, if we give an operator $x^\alpha \partial^\beta/\partial x^\beta$ the degree $|\beta| - |\alpha|$.

EXAMPLE 12.12 Recall that the Riemann curvature operator R may be regarded as a 2-form with values in $\operatorname{End}(TM)$. Let $X \in \mathcal{X}(M)$ be a given tangent vector field. On $T_p M$, the function $v \mapsto (R_p X_p, v)$ is a linear map $T_p M \to \wedge^2 T_p^*(M)$; identifying T with T^* via the metric we may regard this as a degree one polynomial function on T_p with values in $\wedge^2 T_p$. This member of $\mathfrak{P}(TM) \otimes \wedge^* TM$ constructed by this process will be denoted by $(RX, \cdot)$.

PROPOSITION 12.13 *There is a unique symbol map*

$$\sigma_\bullet \colon \mathfrak{D}(S) \to C^\infty(\mathfrak{P}(TM) \otimes \bigwedge\nolimits^* TM \otimes \operatorname{End}_{\mathrm{Cl}}(S))$$

which has the following effect on generators:

(i) $\sigma_0(F) = F$ *for a Clifford module endomorphism F;*
(ii) $\sigma_1(c(X)) = e(X)$, *that is exterior multiplication by X, for $X \in \mathcal{X}(M)$;*
(iii) $\sigma_1(\nabla_X) = \partial_X + \frac{1}{4}(RX, \cdot)$, *where the notation is that of 12.12.*

It is clear that such a symbol map is uniquely determined by its effect on the generators. In fact, the specification of its effect on the generators does determine a unique symbol map on $\otimes_B^* V$, where $B = \operatorname{End}_{\mathrm{Cl}}(S)$ and $V = \mathcal{X}(M) \oplus \mathcal{X}(M)$. The question is whether the symbol is well-defined on $\mathfrak{D}(M)$; does it factor through the quotient map $\otimes_B^* V \to \mathfrak{D}(S)$, or in other words does it respect the relations between the chosen generators of $\mathfrak{D}(S)$. We will complete the proof that $\sigma_\bullet$ is well-defined in the next section, by considering the action of a differential operator on a suitable space of formal Taylor series. Here, however, is an example to show that $\sigma_\bullet$ does respect a crucial relation, that which expresses the curvature as the commutator of covariant derivatives.

EXAMPLE 12.14 In $\mathfrak{D}(S)$ we have the relation

$$\nabla_X \nabla_Y - \nabla_Y \nabla_X - \nabla_{[X,Y]} = K(X,Y) = R^S(X,Y) + F^S(X,Y) \qquad (12.15)$$

where we have decomposed the curvature K into the sum of the Riemann endomorphism and the twisting curvature according to proposition 3.16. Let us calculate the second order Getzler symbols σ_2 of both sides of this identity and verify that they agree. On the left, the operator $\nabla_{[X,Y]}$ is of first order and so may be ignored. It is convenient now to work in local coordinates; let e_i be an orthonormal basis of T_pM, with associated coordinate functions x^i. Then[1]

$$\sigma_1(\nabla_i) = \frac{\partial}{\partial x^i} - \tfrac{1}{8}\sum_{j,k,l}(R(e_i,e_j)e_k,e_l)x^j e_k \wedge e_l.$$

Now when we calculate $\sigma_2(\nabla_i \otimes \nabla_j - \nabla_j \otimes \nabla_i) = \sigma_1(\nabla_i)\sigma_1(\nabla_j) - \sigma_1(\nabla_j)\sigma_1(\nabla_i)$, the second order derivative and the 4-form terms will cancel between the two products and we are left with the cross-terms, which are equal and give a total of

$$[\sigma_1(\nabla_i),\sigma_1(\nabla_j)] = \tfrac{1}{4}\sum_{k,l}(R(e_i,e_j)e_k,e_l)e_k \wedge e_l,$$

which is the symbol of the Riemann endomorphism $R^S(e_i,e_j)$. The twisting curvature F^S is a Clifford module endomorphism, and so has degree zero; so we have verified that the second order symbols of the left and right hand sides of equation 12.15 agree.

Continuing for the moment to take on trust the existence of the *Getzler symbol map*

$$\sigma_\bullet : \mathfrak{D}(S) \to C^\infty(\mathfrak{P}(TM) \otimes \bigwedge\nolimits^* TM \otimes \operatorname{End}_{\operatorname{Cl}}(S))$$

defined by 12.13, let us calculate the symbols of some important differential operators.

EXAMPLE 12.16 The Dirac operator D is of Getzler order 2, and its symbol is the exterior derivative operator d_{TM} on each tangent space TM. To see this, we choose a local orthonormal frame and write $D = \sum c(e_i)\nabla_i$; then $\sigma_2(D) = \sum \sigma_1(c(e_i))\sigma_1(\nabla_i)$ We may substitute the symbols of the Clifford multiplication and covariant derivative operators to get

$$\sigma_2(D) = \sum_i e_i \frac{\partial}{\partial x^i} - \tfrac{1}{8}\sum_{ijkl}(R(e_i,e_j)e_k,e_l)x^j e_i \wedge e_k \wedge e_l$$

and the term involving curvature vanishes because of the Bianchi identity.

---

[1] We have used the symmetry $(R(e_i,e_j)e_k,e_l) = (R(e_k,e_l)e_i,e_j)$.

Since $d^2 = 0$, the fourth order symbol of D^2 vanishes. In fact, although this is not obvious, D^2 also has Getzler order 2; the computation of its second order symbol is crucial to the proof of the index theorem.

PROPOSITION 12.17 The operator D^2 has Getzler order 2. Its Getzler symbol relative to an orthonormal basis of T_pM is

$$-\sum_i \left(\frac{\partial}{\partial x^i} + \frac{1}{4}\sum_j R_{ij}x^j\right)^2 + F^S$$

where R_{ij} is the Riemann curvature at p (thought of as a matrix of 2-forms) and F^S is the twisting curvature 2-form at p.

PROOF This follows from the Weitzenbock formula 3.18, which states that

$$D^2 = \nabla^*\nabla + \tfrac{1}{4}\kappa + F^S$$

where F^S is the Clifford contraction of the twisting curvature. The formula 3.9 for the operator ∇^* gives, in local coordinates,

$$\nabla^*\nabla = \sum_{i,j,k} -g^{jk}(\nabla_j\nabla_k - \Gamma^i_{jk}\nabla_i)$$

where the functions Γ^i_{jk} are the Christoffel symbols associated to the Riemannian connection. Notice that at p, the origin of coordinates, $g^{jk} = \delta^{jk}$ and $\Gamma^i_{jk} = 0$; so the second order Getzler symbol of $\nabla^*\nabla$ is the same as that of the operator $-\sum_i \nabla_i\nabla_i$, which is

$$-\sum_i \left(\frac{\partial}{\partial x^i} + \tfrac{1}{4}\sum_j R_{ij}x^j\right)^2.$$

The second order Getzler symbol of F^S is simply F^S, and the second order Getzler symbol of κ is zero, so the result follows. □

The Getzler symbol of the heat kernel

We will apply the Getzler calculus to the asymptotic expansion of the heat kernel of the Dirac operator, which we derived in 7.15:

$$k_t(p,q) \sim \frac{1}{(4\pi t)^{n/2}} \exp\left\{-\frac{d(p,q)^2}{4t}\right\} \sum_j t^j \Theta_j(p,q) \qquad (12.18)$$

with $\Theta_0(p,p)$ equal to the identity. The heat operator is obviously not a differential operator; we will extend the Getzler calculus to deal with smoothing operators of this sort.

DEFINITION 12.19 For a vector space V, let $\mathbb{C}[[V]]$ denote the ring of formal power series over V (formally speaking, this is the infinite direct product $\prod_{i=0}^{\infty} \otimes^i V$).

The algebra $\mathfrak{P}(V)$ of polynomial coefficient differential operators acts naturally on $\mathbb{C}[[V]]$. If we make $\mathbb{C}[[V]]$ into a graded vector space by giving a monomial x^α the degree $-|\alpha|$, this makes $\mathbb{C}[[V]]$ into a graded $\mathfrak{P}(V)$-module.

Now let s be a smooth section of the bundle $S \boxtimes S^*$ on $M \times M$. Fix $q \in M$ and choose geodesic local coordinates x^i with origin q. Taylor's theorem tells us that the function $s_q(x)$ which is the local coordinate representation of $p \mapsto s(p,q)$ can be expanded (asymptotically) near zero in a Taylor series

$$s_p(x) \sim \sum_\alpha s_\alpha x^\alpha$$

where the s_α are synchronous sections of $S \otimes S_q^*$ (that is, they are parallel along geodesics emanating from q). Since each s_α is determined by its value $s_\alpha(0) \in \text{End}(S_q)$, the Taylor series may be thought of as an element of $\mathbb{C}[[T_qM]] \otimes \text{End}\, S_q$. As q varies we obtain a section $\Sigma(s)$ of the bundle $\mathbb{C}[[TM]] \otimes \text{End}\, S$. We will call this section simply the Taylor series of s.

The algebra $\mathbb{C}[[T_qM]] \otimes \text{End}\, S_q$ is filtered; its filtration is the tensor product (exercise 12.30) of the filtration coming from the grading of $\mathbb{C}[[T_qM]]$ and the Clifford filtration of $\text{End}(S_q)$. We will use this filtration to induce a filtration on the space of smooth sections s of $S \boxtimes S^*$; so we say that s has degree $\leq m$ if its Taylor series $\Sigma(s)$ has degree $\leq m$ (in the product filtration) at each point. The Clifford symbol $\text{End}(S_q) \to \bigwedge^* T_qM \otimes \text{End}_{\text{Cl}}(S_q)$, composed with the Taylor expansion map Σ, gives rise to a symbol map

$$\sigma_\bullet \colon C^\infty(S \boxtimes S^*) \to C^\infty(\mathbb{C}[[TM]] \otimes \bigwedge\nolimits^* TM \otimes \text{End}_{\text{Cl}}(S)).$$

DEFINITION 12.20 We will call the degree m of s relative to the filtration that we have defined above its *Getzler degree*, and we will call the symbol $\sigma_m(s)$ the *Getzler* *symbol* of s. We will also use the notation $\sigma_m^0(s)$ for the constant term in the Taylor series $\sigma_m(s)$, which we will call the *constant part* of the Getzler symbol.

REMARK 12.21 The natural product on smoothing kernels $C^\infty(S \boxtimes S^*)$ is, of course, given by the composition of the corresponding smoothing operators. The symbol that we have defined does *not* have the homomorphism-like property with respect to this product (it does have this property with respect to 'pointwise multiplication' of kernels, defined in an appropriate way). But this is irrelevant for our purposes; the relevant algebraic structure is the action of the differential operators $\mathfrak{D}(S)$ on $C^\infty(S \boxtimes S^*)$, and here there is a good relationship between the Getzler symbols of differential operators defined in the previous section, and the newly defined Getzler symbols of smoothing kernels. This is described in the next proposition.

PROPOSITION 12.22 *Let $T \in \mathfrak{D}(S)$ be one of the generators used in the previous section; that is, T is either a Clifford module endomorphism F, a Clifford multiplication operator $c(X)$, or a covariant derivative ∇_X. Let $m \in \{0,1\}$ be the Getzler order of T. Then for any smoothing operator Q on $C^\infty(S)$, with Getzler order $\leqslant k$, the smoothing operator TQ has Getzler order $\leqslant m + k$, and the relation*

$$\sigma_{m+k}(TQ) = \sigma_m(T)\sigma_k(Q)$$

holds between the symbols.

The 'composition' on the right hand side of this inequality is obtained from the module action of $\mathfrak{P}(TM)$ on $\mathbb{C}[[TM]]$ and the algebra structure of $\wedge^* TM$ and $\mathrm{nd}_{\mathrm{Cl}}(TM)$.

PROOF Let Q have smoothing kernel s, fix $q \in M$ and geodesic coordinates x^i with origin q, and let $s_q(x) \sim \sum s_\alpha x^\alpha$ be the Taylor expansion of s near q.

First we consider the case $T = F$, a Clifford module endomorphism. If F happens to be synchronous at q then the Taylor coefficients of Fs are precisely Fs_α, so the result is obvious in this case. In general let F_0 be the synchronous section of $\mathrm{End}_{\mathrm{Cl}}(S)$ which agrees with F at q; then $F - F_0$ has vanishing constant term in its Taylor expansion, so $\sigma_0(F - F_0) = 0$ and thus

$$\sigma_k(Fs) = \sigma_k(F_0 s) = \sigma_0(F_0)\sigma_k(s) = \sigma_0(F)\sigma_k(s)$$

as required.

Second, if $T = c(X)$ where X is a vector field an exactly similar argument shows that $\sigma_{k+1}(c(X)s) = \sigma_1(c(X))\sigma_k(s)$.

Third, and most important, let us consider the case where X is a vector field and $T = \nabla_X$. Let ∂_i be the vector fields associated to the geodesic local coordinate system x^i; it suffices to prove the result for $X = \partial_i$. Let $Y = \sum_j x^j \partial_j$ be the radial vector field.

Suppose that s is a synchronous section; then $\nabla_Y s = 0$. Let

$$\nabla_X s \sim \sum_\alpha t_\alpha x^\alpha$$

be the Taylor expansion of $\nabla_X s$. By definition

$$\nabla_X \nabla_Y s - \nabla_Y \nabla_X s - \nabla_{[X,Y]} s = K(X,Y)s$$

where K is the curvature operator. But $\nabla_Y s = 0$, and an easy calculation shows that $[X, Y] = X$ and $Y \cdot x^\alpha = |\alpha| x^\alpha$, so we get

$$-\sum_\alpha (|\alpha| + 1) t_\alpha x^\alpha \sim K(X,Y)s = \sum_j K_{ij} x^j s.$$

Thus the Taylor coefficients of $\nabla_X s$ are determined by the Taylor coefficients of K. But we have the identity $K = R^S + F^S$, where the Riemann endomorphism R^S is an element of the Clifford algebra of degree $\leqslant 2$, and F^S is a Clifford module endomorphism. Therefore if we retain only the terms of degree $\leqslant k + 1$ in the above expansion we get

$$\nabla_X s = -\tfrac{1}{2} \sum_j x^j R^S(\partial_i, \partial_j) s + \text{lower order terms}.$$

But the second order symbol of $R^S(\partial_i, \partial_j)$ is $-\tfrac{1}{2} R_{ij}$, so that

$$\sigma_{k+1}(\nabla_X s) = \tfrac{1}{4} \sum_j R_{ij} x^j \wedge \sigma_k(s) = \sigma_1(\nabla_X) \sigma_k(s)$$

which verifies the desired identity in case s is synchronous. The general case follows by applying the special case to each of the coefficients s_α appearing in the Taylor expansion. □

This allows us to give our deferred proof of the remaining part of 12.13.

COROLLARY 12.23 *The Getzler symbol is well defined on $\mathfrak{D}(S)$, and satisfies the identity*
$$\sigma_{m+k}(TQ) = \sigma_m(T)\sigma_k(Q)$$
for all $T \in \mathfrak{D}(S)$ of Getzler order $\leqslant m$, and all Q of Getzler order $\leqslant k$.

PROOF Given $T \in \mathfrak{D}(S)$ of Getzler order $\leqslant m$, let $\tilde{T}$ denote a particular representation of T in terms of generators and relations. We must show that the symbol $\sigma_m(\tilde{T})$ depends only on T. But by repeatedly applying 12.22 we see that
$$\sigma_{m+k}(TQ) = \sigma_m(\tilde{T})\sigma_k(Q)$$
and since $\sigma_k(Q)$ may be an arbitrary formal power series the polynomial coefficient differential operator $\sigma_m(\tilde{T})$ is uniquely determined by this equation. □

We apply the calculus that we have developed to the heat kernel $k_t(p,q)$. The heat kernel has the asymptotic expansion given by 12.18, and it satisfies the equation
$$\left[\frac{\partial}{\partial t} + D_p^2\right] k_t(p,q) = 0.$$

PROPOSITION 12.24 *The terms $\Theta_j(p,q)$ is the asymptotic expansion of the heat kernel have Getzler order $\leqslant 2j$. The 'heat symbol'*
$$W_t = h_t(\sigma_0\Theta_0 + t\sigma_2\Theta_1 + \cdots + t^{n/2}\sigma_n\Theta_{n/2})$$
satisfies the equation $\partial W/\partial t + \sigma_2(D^2)W = 0$, and it is the unique solution of this equation of the form $h_t(v_0 + tv_1 + \cdots + t^{n/2}v_{n/2})$ in which v_j is a symbol of Getzler order $2j$ and $v_0 = 1$.

PROOF We recollect from 7.15 the process by which the asymptotic expansion of the heat kernel was constructed. Suppose that the heat kernel is represented, in local coordinates near q, by a formal series $(x,t) \mapsto h_t(x)(u_0(x) + tu_1(x) + \cdots)$. Using the formula for the commutator of D^2 with multiplication by a smooth function, and the fact that h_t approximately satisfies the heat equation, we obtained a system of differential equations
$$\nabla_{\partial/\partial r}\left(r^j g^{1/4} u_j\right) = -r^{j-1} g^{1/4} D^2 u_{j-1}$$

where by convention we put $u_{j-1} = 0$. These equations determine the u_j uniquely from the single requirement that $u_0(0)$ be the identity. Comparing the Taylor series on the left and the right hand sides, and remembering that $\nabla_{\partial/\partial r}$ annihilates synchronous sections, we see by induction that u_j has Getzler order $\leq 2j$ and that

$$\frac{\partial}{\partial r}\left(r^j \sigma_{2j}(u_j)\right) = r^{j-1}\sigma_2(D^2)\sigma_{2j-2}(u_{j-1})$$

on $T_q M$. These however are precisely the recurrence relations satisfied by the asymptotic-expansion coefficients of the solution to the equation $\partial W/\partial t + \sigma_2(D)W = 0$. Since the recurrence relations determine the coefficients uniquely (given the value of u_0 at the origin), the claim follows. □

The exact solution

We will now construct an explicit solution to the differential equation $\partial W/\partial t + \sigma_2(D^2)W = 0$ which appears in proposition 12.24. Because of the uniqueness assertion in that proposition, this will give us an explicit formula for the heat symbol.

PROPOSITION 12.25 *Suppose that $\hat{R}_{ij}$ is a skew symmetric matrix of real scalars, and that $\hat{F}$ is a real scalar. Then the differential equation*

$$\frac{\partial w}{\partial t} - \sum_i \left(\frac{\partial}{\partial x^i} + \tfrac{1}{4}\sum_j \hat{R}_{ij} x^j\right)^2 w + \hat{F} w = 0$$

has a solution for small t which is an analytic function of the matrix entries $\hat{R}_{ij}$ and of $\hat{F}$, and which is asymptotic to $(4\pi t)^{-n/2}\exp(-|x|^2/4t)$ as t approaches 0. Explicitly this solution is equal to

$$(4\pi t)^{-n/2} \det{}^{1/2}\left(\frac{t\hat{R}/2}{\sinh t\hat{R}/2}\right) \exp\left(-\frac{1}{4t}\left\langle\frac{t\hat{R}}{2}\coth\frac{t\hat{R}}{2}x, x\right\rangle\right)\exp(-t\hat{F}).$$

PROOF An obvious substitution shows that it is enough to prove the special case $\hat{F} = 0$. We use separation of variables. There is an orthonormal basis with respect to which the matrix of R is a direct sum of 2×2 blocks

$$\begin{pmatrix} 0 & \theta \\ -\theta & 0 \end{pmatrix}$$

with eigenvalues $\pm i\theta$. It is enough then to prove the 2-dimensional case, that if R has this 2×2 block form, then the heat kernel is

$$w(x,t) = (4\pi t)^{-1} \left(\frac{it\theta/2}{\sinh(it\theta/2)}\right) \exp\left(-\tfrac{1}{8}i\theta|x|^2 \coth(it\theta/2)\right).$$

The differential equation to be solved is $\partial w/\partial t + Lw = 0$, where

$$L = -\left(\frac{\partial}{\partial x} - \frac{\theta y}{4}\right)^2 - \left(\frac{\partial}{\partial y} + \frac{\theta x}{4}\right)^2 = L_0 + L_1$$

with

$$L_0 = -\left(\frac{\partial^2}{\partial x^2} - \frac{\partial^2}{\partial y^2}\right) - \frac{1}{16}\theta^2(x^2 + y^2)$$

$$L_1 = \tfrac{1}{2}\theta\left(x\frac{\partial}{\partial y} - y\frac{\partial}{\partial x}\right).$$

The claimed solution w is invariant under rotations of $\mathbb{R}^2$. Therefore, $L_1 w = 0$, since L_1 is the infinitesimal generator of the rotation group acting on $\mathbb{R}^2$, and thus annihilates rotationally symmetric functions. It will suffice then to show that $\partial w/\partial t + L_0 w = 0$. Now we can separate variables further into x and y. By Mehler's formula (see 9.12 and remark 9.13), a solution to the equation

$$\frac{\partial w}{\partial t} - \frac{\partial^2 w}{\partial x^2} - \frac{1}{16}\theta^2 x^2 w = 0$$

$$(4\pi t)^{-\frac{1}{2}}\left(\frac{it\theta/2}{\sinh it\theta/2}\right)^{\frac{1}{2}} \exp\left(-\tfrac{1}{8}i\theta x^2 \coth(it\theta/2)\right).$$

Taking the product of this formula for x and the corresponding formula for y, we get the result. □

The operator $\sigma_2(D^2)$ is equal to

$$-\sum_i \left(\frac{\partial}{\partial x^i} + \tfrac{1}{4}\sum_j R_{ij}x^j\right)^2 + F$$

as we calculated above. Here the curvature R is a skew symmetric matrix whose entries are 2-forms, and F is a 2-form with values in $\text{End}_{\text{Cl}}(S)$. The matrix entries R all commute with one another and with F.

163

Two-forms are nilpotent elements of the exterior algebra. Therefore, if we think of the formula from 12.25

$$W = (4\pi t)^{-n/2} \det^{1/2}\left(\frac{tR/2}{\sinh tR/2}\right)\exp\left(-\frac{1}{4t}\left\langle\frac{tR}{2}\coth\frac{tR}{2}x,x\right\rangle\right)\exp(-tF)$$

as a formal power series in the entries of R and F, it converges for all values of t, and (by analytic continuation) it gives a solution to the equation $\partial W/\partial t + \sigma_2(D^2)W = 0$. Moreover, by explicit calculation, this solution W has an expansion

$$\frac{1}{(4\pi t)^{n/2}}(v_0 + tv_1 + \cdots + t^{n/2}v_{n/2})$$

where the formal power series v_j has Getzler order $\leq 2j$, and $v_0(0) = 1$. From the uniqueness assertion of proposition 12.24 we therefore obtain in particular:

PROPOSITION 12.26 *The constant parts of the Getzler symbols of the terms appearing in the asymptotic expansion of the heat kernel for the Dirac-type operator D are given by*

$$\sum_{j=0}^{n/2}\sigma^0_{2j}(\Theta_j) = \det^{1/2}\left(\frac{R/2}{\sinh R/2}\right)\exp(-F) \in \overset{*}{\bigwedge}TM \otimes \mathrm{End}_{\mathrm{Cl}}(S).$$

The index theorem

We have now arrived at the main theorem of this book.

ATIYAH-SINGER INDEX THEOREM 12.27 *Let M be a compact, even-dimensional oriented manifold and let S be a canonically graded Clifford bundle over it with associated Dirac operator D. Then*

$$\mathrm{Ind}(D) = \int_M \hat{\mathcal{A}}(TM) \wedge \mathrm{ch}(S/\Delta)$$

where $\mathrm{ch}(S/\Delta)$ denotes the relative Chern character of S as defined in 4.25. In particular, if M is a spin manifold and $S = \Delta$ is the spinor bundle, the index of the Dirac operator on Δ is equal to the $\hat{\mathcal{A}}$-genus of the manifold M.

REMARK 12.28 In the notation of cohomology theory the right-hand side of the formula is usually written $\langle\hat{\mathcal{A}}(TM) \smile \mathrm{ch}(S/\Delta),[M]\rangle$, where $\smile$ denotes the cup product and $[M]$ is the fundamental homology class of M.

REMARK 12.29 It is easy to extend the statement of the theorem to the case of a general grading on S. Recall from 11.3 that S can be written as a direct sum $S_c \oplus S_a$ of canonically and anticanonically graded sub-bundles. Using D_c and D_a to denote the corresponding Dirac operators, we have

$$\mathrm{Ind}(D) = \mathrm{Ind}(D_c) - \mathrm{Ind}(D_a) = \int_M \hat{A}(TM) \wedge \mathrm{ch}_s(S/\Delta),$$

where the relative super Chern character is defined by $\mathrm{ch}_s(S/\Delta) = \mathrm{ch}(S_c/\Delta) - \mathrm{ch}(S_a/\Delta)$.

PROOF OF THE INDEX THEOREM Recall from the previous chapter (11.14) that

$$\mathrm{Ind}(D) = \frac{1}{(4\pi)^{n/2}} \int_M \mathrm{tr}_s \Theta_{n/2}$$

where $\Theta_{n/2}$ is one of the asymptotic-expansion coefficients that we have been considering. Moreover by Lemma 11.5, the supertrace of $\Theta_{n/2}$, which belongs to $\mathrm{Cl}(TM) \otimes \mathrm{End}_{\mathrm{Cl}}(S)$, can be computed from the top degree part of $\Theta_{n/2}$ in the filtration of the Clifford algebra. Now the symbol $\sigma_n^0(\Theta_{n/2})$ picks out this top degree part, and so by 11.4 and 11.5,

$$\mathrm{tr}_s \Theta_{n/2} = (-2i)^{n/2} \mathrm{tr}^{S/\Delta}(\sigma_n^0(\Theta_{n/2})).$$

However by 12.26, the symbol $\sigma_n^0(\Theta_{n/2})$ is exactly equal to the n-form part of $\det^{1/2}\left(\frac{R/2}{\sinh R/2}\right)\exp(-F)$, and therefore $\mathrm{tr}\,\Theta_{n/2}$ is the n-form part of

$$(-2i)^{n/2} \det^{1/2}\left(\frac{R/2}{\sinh R/2}\right) \mathrm{tr}^{S/\Delta}(\exp(-F)).$$

But by definition of the $\hat{A}$-genus and of the relative Chern character, this equals $-2i)^{n/2}(+2\pi i)^{n/2}$ times the n-form part of $\hat{A}(TM)\mathrm{ch}(S/\Delta)$. The index theorem follows. □

Notes

Getzler's proof of the index theorem appeared in [31], with a different version in [32]. The argument in this chapter is rather closer to that in [31], although the reader should note that the symbol algebra defined in [31] is the 'Fourier transform' of that which we have employed here. For the 'rescaling' approach of [32], see exercise 12.31.

The 'Getzler calculus' has been employed in many other index-theoretic calculations involving the asymptotics of Dirac-type operators; for a significant example see [23].

A number of similar proofs of the index theorem have appeared; for bibliography and discussion consult [12], especially the notes at the end of Chapter 4.

The original statement of the Index Theorem applied to any elliptic operator on a manifold, not necessarily of generalized Dirac type. It can be shown however that any such operator is equivalent, for index-theoretic purposes, to a generalized Dirac operator. This assertion is implicit in [9]; the fundamental reason is the appearance of Dirac operators in the formulae for Poincaré duality between K-theory and K-homology.

Exercises

QUESTION 12.30 Let A and B be filtered algebras. Show that a filtration on $A \otimes B$ may be defined by
$$(A \otimes B)_m = \sum_{k+l=m} A_k \otimes B_l.$$
(This is called the *tensor product filtration*.)

QUESTION 12.31 Let M be a compact spin manifold, S the spinor bundle, and for a point $q \in M$ let $\mathcal{O}_q(S)$ denote the space of germs of sections of $S \otimes S_q^*$ near to q. By choosing geodesic coordinates with q as origin, and trivializing S near q by radial parallel transport, we may identify elements of $\mathcal{O}_q(S)$ with germs of smooth functions $f \colon \mathbb{R}^n \to \mathrm{End}(S_q) = \mathrm{Cl}(\mathbb{R}^n)$. We will write such an f as $\sum_{k=0}^n f_k(x)$, where $f_k \in \mathrm{Cl}(\mathbb{R}^n)^k \ominus \mathrm{Cl}(\mathbb{R}^n)^{k-1}$. *Getzler's rescaling* is the map $R_\lambda \colon \mathcal{O}_q \to \mathcal{O}_q$ defined by
$$R_\lambda f(x) = \sum_{k=0}^n \lambda^{n-k} f_k(\lambda x).$$
Show that if $D \in \mathfrak{D}(S)$ has Getzler order $\leqslant m$, then
$$\sigma_m(D) = \lim_{\lambda \to 0} \lambda^m R_\lambda^{-1} D R_\lambda$$
in an appropriate topology. (Notice that this gives a different proof that the Getzler symbol of a differential operator is well-defined.)

QUESTION 12.32 Extending the argument of the previous question, suppose that f is now the germ of a *time-dependent* section of $S \otimes S_q^*$, and define

$$R_\lambda f(x,t) = \sum_{k=0}^{n} \lambda^{n-k} f_k(\lambda x, \lambda^2 t).$$

If f is a solution to the heat equation $\partial f/\partial t + D^2 f = 0$, show that $g = R_\lambda^{-1} f$ is a solution to the rescaled equation $(\partial/\partial t + R_\lambda^{-1} D^2 R_\lambda)g = 0$. Now deduce the index theorem, by letting $\lambda \to 0$ and making use of the fact that the asymptotic expansion coefficients for $\partial/\partial t + L$ depend continuously on the coefficients of the operator L. See [32, appendix B].)

CHAPTER 13

Applications of the index theorem

In this chapter we will review a number of classical applications of the Index Theorem. Some of the results (such as Hirzebruch's signature theorem) actually predate the index theorem itself, and were instrumental as motivation for the first proof. Others (such as the index theorem for the spinor Dirac operator) were among the first new consequences to flow from it. We begin with the spinor Dirac case.

The spinor Dirac operator

Let M be a compact even-dimensional spin-manifold, Δ the associated spin bundle, and D the Dirac operator on Δ. In this case the index theorem takes the form

$$\text{Ind}(D) = \langle \widehat{\mathcal{A}}(M), [M] \rangle.$$

Recall that the $\widehat{\mathcal{A}}$-genus of M which appears on the right hand side is a certain combination of the Pontrjagin classes of TM. In fact we have

$$\widehat{\mathcal{A}}_4 = -p_1/24, \quad \widehat{\mathcal{A}}_8 = (-4p_2 + 7p_1^2)/5760, \quad \widehat{\mathcal{A}}_{12} = (-16p_3 + 44p_1 p_2 - 31p_1^3)/967680$$

where $\widehat{\mathcal{A}}_n$ denotes n-dimensional component of $\widehat{\mathcal{A}}$. We recall from lemma 2.27 how these expressions are calculated. Let

$$g(z) = \frac{\sqrt{z}/2}{\sinh \sqrt{z}/2} = 1 - \frac{1}{24}z + \frac{7}{5760}z^2 + \cdots.$$

Expand the product $\prod g(y_j)$, where the y_j are formal variables, as a symmetric formal power series, and then substitute the Pontrjagin class p_i for the i'th elementary symmetric function in the formal variables y_j. The result is the expansion of the $\widehat{\mathcal{A}}$-class in terms of the Pontrjagin classes.

An early application of the spinor index theorem was to the study of topological obstructions to positive scalar curvature.

THEOREM 13.1 (LICHNEROWICZ [48]) *Let M be a compact manifold which admits a spin structure, and for which the $\widehat{A}$-genus $\langle \widehat{A}(M), [M]\rangle$ is non-zero. Then M admits no metric of strictly positive scalar curvature.*

PROOF This is a consequence of the Bochner vanishing argument (3.10). Note that for the spinor Dirac operator, the Weitzenbock formula just says

$$D^2 = \nabla^*\nabla + \tfrac{1}{4}\kappa$$

where κ is the scalar curvature. (This follows from 3.18 and 4.21). Thus, if $\kappa > 0$, the Bochner argument shows that the kernel $\ker D = \ker D^2$ is zero. But then $\mathrm{Ind}(D) = \widehat{A}(M)$ is zero also, a contradiction. □

REMARK 13.2 In high dimensions the scalar curvature is a very weak invariant of the geometry of a manifold, since it is determined by averaging a large number of components of the Riemann curvature tensor. This means that it is difficult to control the possible scalar curvatures of metrics on a compact manifold M; in fact, it is known that there is no obstruction to any manifold having a metric of *negative* scalar curvature[1]. Thus the simple obstruction to *positive* scalar curvature provided by Lichnerowicz' theorem is a remarkable one.

There is in fact a well-developed theory of positive scalar curvature manifolds, in which Lichnerowicz' theorem appears as the first of a series of obstructions which are related to the cohomology of the fundamental group. To investigate the higher obstructions one needs a version of the Index Theorem which works on the universal cover of a compact manifold, taking the fundamental group action into account. Such a theory can be developed using the K-theory of operator algebras. In chapter 15 we will discuss (in an elementary way) a simple example of a higher index theorem of this sort.

One piece of information which the Index Theorem immediately implies is that the $\widehat{A}$-genus of a spin manifold is an *integer*. The expression of the $\widehat{A}$-genus in terms of the Pontrjagin classes does not provide any a priori reason for this; in fact, the $\widehat{A}$-genus of a non-spin manifold need not be an integer (exercise 13.18), and Atiyah

---

[1] Or even of negative *Ricci* curvature! see [49].

has recorded that the question 'why is the $\hat{A}$-genus of a spin manifold an integer?' provided one of the original motivations for the Index Theorem. In certain dimensions one can get slightly more information by studying the real (as opposed to complex) representation theory of the Clifford algebra.

PROPOSITION 13.3 *The $\hat{A}$-genus of a 4-dimensional spin manifold is an even integer.*

In fact, the conclusion holds in any dimension congruent to 4 modulo 8 (see [47]) but we do not propose to develop the theory needed to prove the result in this generality.

PROOF We need to know about the structure of the four-dimensional Clifford algebra. Recall that the skew-field $\mathbb{H}$ of *quaternions* is spanned as an $\mathbb{R}$-vector space by 1, i, j, and k, with $i^2 = j^2 = k^2 = ijk = -1$. Now consider the matrix algebra $M_2(\mathbb{H})$. The matrices

$$e_1 = \begin{pmatrix} 0 & i \\ i & 0 \end{pmatrix}, \quad e_2 = \begin{pmatrix} 0 & j \\ j & 0 \end{pmatrix}, \quad e_3 = \begin{pmatrix} 0 & k \\ k & 0 \end{pmatrix}, \quad e_4 = \begin{pmatrix} 0 & 1 \\ -1 & 0 \end{pmatrix}$$

all have square -1 and anticommute, so they generate a homomorphism of algebras $\mathrm{Cl}(\mathbb{R}^4) \to M_2(\mathbb{H})$, which by dimension counting must be an isomorphism. Thus $\mathrm{Cl}(\mathbb{R}^4)$ acts as a matrix algebra on the two-dimensional quaternionic (right) vector space $\mathbb{H}^2$, and therefore $\mathrm{Cl}(\mathbb{R}^4) \otimes \mathbb{C}$ acts as a matrix algebra on the underlying complex vector space $\mathbb{C}^4$, which (again by dimension counting) is just the spin representation. The point of this calculation is to show that the spin representation of $\mathrm{Cl}(\mathbb{R}^4)$ has a natural quaternionic structure; it is the complex vector space underlying a quaternionic vector space, or, equivalently, it is provided with a canonical antilinear anti-involution J (equal to right multiplication by the quaternion j). Therefore the spin representation of $\mathrm{Spin}(4)$ also has such a quaternionic structure. The spinor bundle Δ can be considered as a bundle of quaternionic vector spaces, and since the quaternionic structure is compatible with the connection and commutes with the Clifford action, the kernels of D_+ and D_- are quaternionic vector spaces also. The result follows since the dimension over $\mathbb{C}$ of a quaternionic vector space must be even.

The signature theorem

Let M be a smooth, oriented, compact Riemannian manifold, of even dimension $2m$. Let D denote the de Rham operator on differential forms on M, i.e. the Dirac operator of $\Lambda^*T^*M \otimes \mathbb{C}$ considered as a Clifford bundle. We will equip this bundle with the canonical grading, in the sense of 11.3; it is defined by the Clifford action of $i^m\omega$, where ω is the volume form in the Clifford algebra. We refer to the associated Dirac operator D (with this grading) as the *signature operator*.

Clifford multiplication by ω is just the Hodge $*$-operator (1.21) up to sign. Thus we may define the grading operator

$$\varepsilon = i^m\omega = i^{m+p(p-1)} * \quad \text{(on p-forms)}$$

without reference to Clifford algebras.

We can evaluate the index of the signature operator in terms of algebraic topology. Suppose that the dimension $2m$ of M is a multiple of 4. Then the cup-product in cohomology induces a symmetric bilinear form (the *intersection form*, see 6.5) on $H^m(M;\mathbb{R})$:

$$H^m(M;\mathbb{R}) \otimes H^m(M;\mathbb{R}) \to H^n(M;\mathbb{R}) \xrightarrow{\int} \mathbb{R}.$$

This bilinear form is non-degenerate, by Poincaré duality (6.4).

DEFINITION 13.4 The *signature* of the $2m$-dimensional oriented manifold M (where m is even) is the signature (that is, the number of positive eigenvalues minus the number of negative eigenvalues) of the intersection form on $H^m(M;\mathbb{R})$.

By its definition, the signature is an invariant of the oriented homotopy type of M

PROPOSITION 13.5 *Let M be a compact oriented manifold of dimension $2m$, where m is even. Then the index of the signature operator on M is equal to the signature of M as defined above.*

PROOF Let us write the Laplacian $\Delta = D^2$ as a direct sum $\Delta^+ \oplus \Delta^-$ relative to the canonical grading ε. Then

$$\text{Ind}(D) = \dim \text{Ker}(\Delta^+) - \dim \text{Ker}(\Delta^-) .$$

Now let Δ_l^+, Δ_l^- denote the restrictions of Δ^+, Δ^- to the ε-invariant subspaces $C^\infty(\Lambda^l T^*M \oplus \Lambda^{2m-l}T^*M)$, $0 \leq l < m$ and $C^\infty(\Lambda^m T^*M)$ ($l = m$). If $l < m$ and $\alpha \in \text{Ker}(\Delta_l^+)$, then $\alpha = \beta + \varepsilon(\beta)$, where β is a harmonic l-form; and then $\beta - \varepsilon(\beta) \in \text{Ker}(\Delta_l^-)$. Therefore, $\text{Ker}(\Delta_l^+)$ and $\text{Ker}(\Delta_l^-)$ are isomorphic for $l < m$, and

$$\begin{aligned}\text{Ind}(D) &= \dim \text{Ker}(\Delta_m^+) - \dim \text{Ker}(\Delta_m^-) \\ &= \dim(\mathcal{H}^+) - \dim(\mathcal{H}^-)\end{aligned}$$

where $\mathcal{H}^+$ and $\mathcal{H}^-$ are the ± 1-eigenspaces of $*$ on harmonic m-forms (notice that $\varepsilon = *$ on m-forms). The quadratic form

$$\alpha \mapsto \int \alpha \wedge \alpha$$

is positive definite on $\mathcal{H}^+$ and negative definite on $\mathcal{H}^-$, so $\text{Ind}(D)$ equals the signature of this quadratic form on the space of harmonic m-forms. The Hodge theorem, 6.2, now completes the proof. □

Now we will calculate the index of the signature operator. Let $S = \Lambda^* T^*M$ be the graded Clifford bundle on which the signature operator acts. We need to know the relative Chern character $\text{ch}(S/\Delta)$.

LEMMA 13.6 *The relative Chern character $\text{ch}(S/\Delta)$ is equal to $2^m \mathfrak{S}(TM)$, where $\mathfrak{S}$ is the Pontrjagin genus associated to the holomorphic function $z \mapsto \cosh(\frac{1}{2}z)$.*

PROOF The bundle S is isomorphic, as a bundle of Clifford modules, to the Clifford algebra $\text{Cl}(TM)$ with its canonical grading. But, locally, $\text{Cl} = \Delta \otimes \Delta^*$ and therefore the relative Chern character of S is equal (locally, as a differential form) to the absolute Chern character of Δ^*, the dual of the spin representation. Proposition 4.23 then identifies this Chern character as $2^m \mathfrak{S}(TM)$. Because these calculations can all be thought of as local ones with the curvature tensor, they remain valid even in the absence of a global spin structure on the manifold M. □

Recall that Hirzebruch's $\mathcal{L}$ class is the Pontrjagin genus associated to the holomorphic function $z \mapsto \sqrt{z}/\tanh\sqrt{z}$.

HIRZEBRUCH SIGNATURE THEOREM 13.7 The signature of a manifold M (of dimension $2m$ divisible by 4) is given by evaluating the $\mathcal{L}$ class on the fundamental homology class; in symbols

$$\text{Sign}(M) = \langle \mathcal{L}(TM), [M] \rangle.$$

PROOF As we have seen, the signature of M is equal to the index of the signature operator D. By the Index Theorem,

$$\text{Ind}(D) = 2^m \langle \hat{\mathcal{A}}(TM)\mathfrak{S}(TM), [M] \rangle.$$

Now $\mathcal{L}_1(TM) = \hat{\mathcal{A}}(TM)\mathfrak{S}(TM)$ is the Pontrjagin genus associated to the holomorphic function

$$g_1(z) = \frac{\sqrt{z}/2}{\sinh(\sqrt{z}/2)} \cdot \{\cosh(\sqrt{z}/2)\} = \frac{\sqrt{z}/2}{\tanh(\sqrt{z}/2)}$$

whereas the $\mathcal{L}$ class is by definition associated to the holomorphic function $g(z) = \sqrt{z}/\tanh\sqrt{z}$. Therefore we have the equality

$$[\mathcal{L}_1(TM)]_k = 2^{-k/4}[\mathcal{L}(TM)]_k$$

between the k-dimensional pieces of $\mathcal{L}_1$ and $\mathcal{L}$; so that in particular $2^m[\mathcal{L}_1(TM)]_{2m} = [\mathcal{L}(TM)]_{2m}$ and the index theorem follows. □

REMARK 13.8 One sees from the proof above that the genus $\mathcal{L}_1$ is in some respects more natural than $\mathcal{L}$, and some authors therefore *define* the $\mathcal{L}$-genus to be what we have called $\mathcal{L}_1$. We have retained the original definition of Hirzebruch.

By calculations analogous to those carried out above for $\hat{\mathcal{A}}$, one can work out the first few components of the $\mathcal{L}$-genus,

$$\mathcal{L}_4 = p_1/3, \quad \mathcal{L}_8 = (7p_2 - p_1^2)/45, \quad \mathcal{L}_{12} = (62p_3 - 13p_1p_2 + 2p_1^3)/945.$$

In particular we see that in dimension four, the signature is equal to -8 times the $\hat{\mathcal{A}}$-genus. From 13.3 we therefore obtain

ROCHLIN'S THEOREM 13.9 The signature of a (smooth) spin four-manifold is divisible by 16.

We emphasized smoothness here for the following reason. The spin condition on a four-manifold has a simple interpretation in terms of the intersection form; an oriented four-manifold is spin if and only if its intersection form is *even*, that is, if its matrix (relative to an integral basis of $H^2(M;\mathbb{Z})$) has its diagonal entries even. It is known for number-theoretic reasons that any even, unimodular quadratic form over $\mathbb{Z}$ must have signature equal to a multiple of 8. The additional factor of 2 in Rochlin's theorem depends crucially upon smoothness; Freedman produced an example of a compact *topological* four-manifold which has even intersection form and signature 8, an indication of the dramatic difference between the topological and smooth categories in this dimension [30, 27].

The Hirzebruch-Riemann-Roch theorem

Now we will briefly discuss the most famous application of the index theorem in complex geometry. Let M be a compact n-dimensional complex manifold. As we saw in 3.25, the complex structure gives an operator J on each real tangent space to M, with $J^2 = -1$, and we get a decomposition of $TM \otimes \mathbb{C}$ into two complex conjugate pieces

$$TM \otimes \mathbb{C} = T^{1,0}M \oplus T^{0,1}M \tag{13.10}$$

where $T^{1,0}M$ is isomorphic as a *complex* bundle to TM. Then (3.26) the bundle $S = \Lambda^*(T^{0,1}M)^*$ carries a spin representation of the bundle of Clifford algebras $\mathrm{Cl}(TM)$. By remark 4.30, M can be endowed with a Spin^c-structure of which S is the spin representation; and the fundamental line bundle for this Spin^c structure is

$$L = \mathrm{Hom}_{\mathrm{Cl}}(\overline{S}, S) = \mathrm{Hom}_{\mathrm{Cl}}(\overset{*}{\Lambda}(T^{1,0}M)^*, \overset{*}{\Lambda}(T^{0,1}M)^*).$$

A homomorphism of Clifford modules from $\Lambda^*(T^{1,0}M)^*$ to $\Lambda^*(T^{0,1}M)^*$ is determined by the image of $1 \in \Lambda^0$, and a moments thought shows that 1 must be mapped to an element of the top exterior power $\Lambda^n(T^{0,1}M)^*$. Thus $L \cong \Lambda^n(T^{0,1}M)^*$.

LEMMA 13.11 *For the Clifford bundle S defined above, the relative Chern character $\mathrm{ch}(S/\Delta)$ is equal to the Chern genus of the complex tangent bundle $T^{1,0}M$ associated to the holomorphic function $z \mapsto e^{-z/2}$.*

PROOF By exercise 2.37, $c_1(L) = c_1((T^{0,1}M)^*) = c_1(T^{1,0}M)$. Therefore the Chern character of the bundle L is e^{-c_1}. However, by 4.29, the twisting curvature of S is half the curvature of L and thus the relative Chern character of S is $e^{-c_1/2}$. This is exactly the Chern genus associated with the function $e^{-z/2}$. □

In applying the Atiyah-Singer index theorem it is helpful to reformulate the expression for the $\hat{A}$-genus. Recall that the Pontrjagin classes of TM are, by definition, the Chern classes of its complexification $TM \otimes \mathbb{C}$. Using the decomposition 13.10 above, and remembering that $c_i(T^{0,1}M) = (-1)^i c_i(T^{1,0}M)$, we can express the Pontrjagin classes of TM in terms of the Chern classes of the complex tangent bundle $T^{1,0}M$. A simple calculation gives

LEMMA 13.12 *The $\hat{A}$-genus of a complex manifold is equal to the Chern genus of its complex tangent bundle associated to the holomorphic function*

$$z \mapsto \frac{z/2}{\sinh z/2}.$$

Let W be a holomorphic vector bundle over M. The space of holomorphic sections of W is then finite-dimensional and in many situations in complex geometry one wants to compute, or at least estimate, its dimension. One may form the *Dolbeault complex* of W

$$\Omega^{0,0}(W) \xrightarrow{\bar{\partial}} \Omega^{0,1}(W) \xrightarrow{\bar{\partial}} \cdots \xrightarrow{\bar{\partial}} \Omega^{0,n}(W)$$

where $\Omega^{0,k}(W)$ denotes the space of sections of the bundle $\bigwedge^k (T^{0,1}M)^* \otimes W$. By Hodge theory this complex has finite-dimensional cohomology groups, of which the first, $H^{0,0}(W)$, is just the space of holomorphic sections of W. The Riemann-Roch theorem computes the Euler characteristic of the Dolbeault complex.

THEOREM 13.13 (HIRZEBRUCH-RIEMANN-ROCH) *In the above situation we have*

$$\sum_k (-1)^k \dim H^{0,k}(W) = \langle \mathrm{Td}(T^{1,0}M) \mathrm{ch}(W), [M] \rangle$$

where the Todd genus Td *of a complex vector bundle is by definition the Chern genus associated to the holomorphic function* $z \mapsto \dfrac{z}{e^z - 1}$.

OUTLINE PROOF Provide W with a hermitian metric and compatible connection. Consider the graded Clifford bundle $S \otimes W$. If M is a Kähler manifold, the Dirac operator of this Clifford bundle is equal to $\sqrt{2}(\bar{\partial} + \bar{\partial}^*)$ (see 3.27), and therefore its index is equal to the Euler characteristic of the Dolbeault complex $\Omega^{0,*}(W)$. In general $\sqrt{2}(\bar{\partial} + \bar{\partial}^*) = D + A$, where $A \in \text{End}(S)$ is a zero order term. The homotopy $D + tA$, $t \in [0,1]$, together with the homotopy invariance of the index (11.13), shows that the index of D is still equal to the Euler characteristic in this case. Now by the Index Theorem,

$$\text{Ind}(D) = \langle \hat{\mathcal{A}}(TM)\,\text{ch}(S/\Delta)\,\text{ch}(W), [M] \rangle$$

and the two lemmas above show that $\hat{\mathcal{A}}(TM)\,\text{ch}(S/\Delta)$ is the Chern genus associated to the holomorphic function

$$\frac{z/2}{\sinh z/2} \cdot e^{-z/2} = \frac{z}{e^z - 1}. \quad \square$$

Local index theory

The classical examples that we have presented so far all depend on the global formula for the index in terms of characteristic classes. However, it is one of the virtues of the heat equation approach to the index theorem that it does not merely identify the index in global, topological terms (as a characteristic class), but also in local, geometrical terms (as a specific differential form). In this final section we will mention without proofs some results that make essential use of this local structure.

As one might expect, locality becomes important on non-compact manifolds.

PROPOSITION 13.14 *Let M be a complete Riemannian manifold, D a Dirac operator on a Clifford bundle S, and suppose that the curvature term $\mathbf{K}$ in the Weitzenbock formula is uniformly positive outside a compact subset of M. Then the L^2-kernel of D is finite-dimensional, and D is invertible on the orthocomplement of this kernel in $L^2(S)$.*

This proposition means that, if S is graded, the index $\text{Ind}(D)$ can be defined as in the compact manifold case. Can we find a formula for the index? Suppose for instance that we are considering the classical Dirac operator on a spin manifold M with a *cylindrical end*. This means that M is the union of two pieces

$$M = M_0 \cup_{\partial M_0 = N} N \times [0, \infty)$$

a compact manifold M_0 with boundary $\partial M_0 = N$, and a semi-infinite cylinder (with the product metric) $N \times [0, \infty)$. If we assume that N has positive scalar curvature, then the conditions of the proposition above are satisfied. Moreover the $\hat{A}$-genus form vanishes along the end, so is compactly supported in M_0. Thus the integral $\int_M \hat{A}$ makes sense, and one might conjecture that this integral equals the index of D.

Examples show that this conjecture is *false*. To quantify its failure, Atiyah, Patodi and Singer introduced the *eta invariant* of the manifold M. Recall from exercise 8.23 that we can define the zeta function associated to a Dirac operator D on a compact manifold such as N by

$$\zeta(s) = \sum_j |\lambda_j|^{-2s}$$

where the numbers λ_j are the non-zero eigenvalues of D. Similarly, we define the eta function by

$$\eta(s) = \sum_j (\operatorname{sgn} \lambda_j)|\lambda_j|^{-2s}$$

where $\operatorname{sgn} \lambda_j \in \{\pm 1\}$ is the sign of λ_j. These Dirichlet series converge for large values of $\Re s$, but using the asymptotic expansion for the heat equation one can show that they can in fact be analytically continued to meromorphic functions on the whole complex plane. It turns out that the eta-function has no pole at zero (this is quite a deep result, proved by a Getzler symbol argument analogous to that used in the proof of the Index Theorem). The value $\eta(0)$ can be thought of as a renormalization of the 'signature' of the quadratic form associated to D, the 'difference between the dimensions' of the (infinite-dimensional!) positive and negative eigenspaces of the operator D. For this reason Atiyah, Patodi, and Singer describe it as a measure of *spectral asymmetry*.

THEOREM 13.15 (ATIYAH-PATODI-SINGER [5]) *Let M be a spin manifold with cylindrical end of positive scalar curvature, as above. Then*

$$\operatorname{Ind} D_M = \int_M \hat{A}(TM) - \tfrac{1}{2}\eta_N(0)$$

where D_M denotes the Dirac operator on M, and η_N is the eta-function associated to the Dirac operator on N.

The proof uses the heat equation method, with a careful analysis of the heat kernel obtained by grafting the construction of chapter 7 on the compact piece M_0 to a construction using separation of variables on the cylindrical end.

REMARK 13.16 Notice that, as a consequence of this result, if M and M' are two different manifolds which have isomorphic cylindrical ends, then the difference of the indices of the Dirac operators on M_0 and M_1 is equal to the difference of the integrals of the $\hat{A}$-forms. A statement of this kind in fact holds in great generality for *any* two manifolds which are 'isomorphic at infinity': this is the *relative index theorem* of Gromov and Lawson [38], a key tool in some studies of the positive scalar curvature problem.

Notes

For a survey of the theory of positive scalar curvature metrics, see [70].

The original reference on the signature theorem is the book by Hirzebruch [40]; this also contains the first version of the generalized Riemann-Roch theorem. The treatment of the signature theorem by Milnor and Stasheff [55] is another classic. In particular this book gives the application of the signature theorem to the construction of (some) 'exotic spheres', that is smooth manifolds homeomorphic, but not diffeomorphic, to the standard sphere. The connections between signatures and the topology of manifolds lead to *surgery theory*, which involves 'inverting' the signature theorem in a certain sense. See [17].

For much more about Riemann-Roch theorems and complex geometry, see [36].

The papers [5, 6, 7] are the original ones on the eta-invariant. The book [52] embeds the theorem in a sophisticated geometric-analytic framework which also includes the proof of the ordinary index theorem.

Higher index theory, that is index theory taking into account the fundamental group or other 'large scale' structure, is a rapidly developing subject. An overview is attempted in [63], and see also [43, 22] for related deep discussions. The transition from 'lower' to 'higher' indices corresponds in surgery theory to the transition from [] to [72].

Exercises

QUESTION 13.17 Show directly (that is, without appealing to the index theorem) that the index of the spinor Dirac operator on a 6-dimensional spin manifold is zero. (You will need to investigate the real structure of the Clifford algebra, as in our proof of Rochlin's theorem.) Can you extend the argument to cover all dimensions congruent to 2 modulo 4?

QUESTION 13.18 Compute the $\hat{A}$-class and the $\mathcal{L}$-class of $\mathbb{CP}^2$. Verify the signature theorem in this case, and show that $\mathbb{CP}^2$ has no spin structure.

QUESTION 13.19 Construct a natural homomorphism $U(k) \to \mathrm{Spin}^c(2k)$ which makes the diagram

$$\begin{array}{ccc} U(k) & \longrightarrow & \mathrm{Spin}^c(2k) \\ & \searrow & \downarrow \\ & & SO(2k) \times U(1) \end{array}$$

commute. (Here the map $U(k) \to SO(2k) \times U(1)$ is equal to $i \times \det$, where $i\colon U(k) \to SO(2k)$ is the natural inclusion.) Hence get another proof that a complex manifold has a natural Spin^c structure, for which the canonical line bundle is the determinant bundle of the complex tangent bundle.

Verify that this is the same Spin^c structure as we used in the text.

QUESTION 13.20 Let M be a compact oriented manifold and let $S = \wedge^* T^*M \otimes$ equipped with the Euler grading (see 11.8). Show that the canonically and anticanonically graded parts of S are locally isomorphic to $\Delta \otimes \Delta_+$ and $\Delta \otimes \Delta_-$ respectively. Using exercise 4.34, obtain the *Chern-Gauss-Bonnet* theorem

$$\sum_i (-1)^i \dim(H^i(M;\mathbb{R})) = \langle e(TM), [M] \rangle$$

from the index theorem.

QUESTION 13.21 Let M be a compact oriented 4-manifold. The *anti-self-dual* (ASD) complex of M is

$$\Omega^0(M) \xrightarrow{d} \Omega^1(M) \xrightarrow{d} \Omega^2_-(M)$$

where Ω^2_- consists of those two-forms for which $\alpha = -*\alpha$. Show that the ASD complex has finite-dimensional cohomology, and compute its Euler characteristic.

(Nonlinear equations involving self-duality are of critical importance in the study of smooth four-manifolds; see [27].)

CHAPTER 14

Witten's approach to Morse theory

Let M be a compact smooth manifold, $h\colon M \to \mathbb{R}$ a suitable smooth function, and for $c \in \mathbb{R}$ let $M_c = \{p \in M : h(p) \leqslant c\}$. For c sufficiently small, $M_c = \emptyset$, and for c sufficiently large, $M_c = M$. The idea on which classical Morse theory depends is that, as c varies, the topology of M_c will not change except when c passes through a critical value of h; and that when c does pass through such a critical value, the change in the topology can be investigated locally, near to the corresponding critical point (or points) of h. Thus the critical point structure of h will give rise to a combinatorial model for the topology of M. For an account of this classical and powerful theory see [54].

In 1982, Witten [73] gave a new approach to some of the ideas of Morse theory. His method was to deform the de Rham complex of M, in a manner depending on h, so that the low-energy eigenvectors of the Laplace operators became concentrated near the critical points of h. The object of this chapter is to give an elementary exposition of some of Witten's argument. For a much more sophisticated discussion see [39].

The Morse inequalities

In Chapter 6 we defined a *Dirac complex* over a compact Riemannian manifold and we proved the Hodge theorem, that the cohomology of such a complex is represented by harmonic sections. The index of the associated Dirac operator is just the Euler characteristic of the complex, that is the alternating sum of the dimensions of the various cohomology groups. If we define the *Betti numbers* of the Dirac complex (S, d) by

$$\beta_j = \dim H^j(S, d), \qquad (14.1)$$

then $\mathrm{Ind}(d + d^*) = \sum (-1)^j \beta_j$. The *Morse inequalities* are a system of inequalities that allow one to estimate the individual Betti numbers β_j.

In analysis, the Morse inequalities arise as follows. Suppose that φ is a smooth rapidly decreasing positive function on $\mathbb{R}^+$ with $\varphi(0) = 1$. Then the operator $\varphi(D^2)$ (where D is the Dirac operator) is smoothing and therefore of trace class. Set

$$\mu_j = \operatorname{Tr}\left(\varphi(D^2)\big|_{S_j}\right). \tag{14.2}$$

Then

PROPOSITION 14.3 *With the hypotheses above, the numbers (μ_j) and (β_j) satisfy the following system of inequalities (known as the Morse inequalities):*

$$\mu_0 \geq \beta_0$$
$$\mu_1 - \mu_0 \geq \beta_1 - \beta_0$$
$$\mu_2 - \mu_1 + \mu_0 \geq \beta_2 - \beta_1 + \beta_0$$

and so on, and finally an equality

$$\sum (-1)^j \mu_j = \sum (-1)^j \beta_j.$$

PROOF By the Hodge theorem (6.2), β_j is equal to the dimension of the kernel of D^2 on sections of S_j. Since the spectrum of D^2 is discrete, there is a smooth function $\tilde{\varphi}$ on $\mathbb{R}^+$ which is positive, rapidly decreasing, with $\tilde{\varphi}(0) = 1$ and $\tilde{\varphi}(\lambda) = 0$ for all non-zero eigenvalues λ of D^2; there is no loss of generality in assuming also that $\tilde{\varphi} \leq \varphi$. Then $\beta_j = \operatorname{Tr}\left(\tilde{\varphi}(D^2)\big|_{S_j}\right)$, so that $\mu_j - \beta_j = \operatorname{Tr}\left((\varphi - \tilde{\varphi})(D^2)\big|_{S_j}\right)$.

We may write the function $\varphi - \tilde{\varphi}$ in the form

$$(\varphi - \tilde{\varphi})(\lambda) = \lambda(\psi(\lambda))^2$$

where ψ is positive and rapidly decreasing, vanishes at zero and is differentiable there. So we may write $(\varphi - \tilde{\varphi})(D^2) = D^2(\psi(D^2))^2$. Now we make a trace argument exactly as in the proof of 10.7. We have $D^2 = dd^* + d^*d$ and

$$\begin{aligned}
\operatorname{Tr}\left(dd^*(\psi(D^2))^2\big|_{S_j}\right) &= \operatorname{Tr}\left(\psi(D^2)\big|_{S_j} dd^*\psi(D^2)\big|_{S_j}\right) \\
&= \operatorname{Tr}\left(d^*(\psi(D^2))^2\big|_{S_j} d\right) \\
&= \operatorname{Tr}\left(d^*d(\psi(D^2))^2\big|_{S_{j-1}}\right).
\end{aligned}$$

Therefore
$$(\mu_j - \beta_j) - (\mu_{j-1} - \beta_{j-1}) + (\mu_{j-2}\,\beta_{j-2}) - \cdots = \text{Tr}\left(d^*d(\psi(D^2))^2\big|_{S_j}\right).$$
If j equals the top dimension of the complex, then this is zero. In general, write
$$d^*d(\psi(D^2))^2\big|_{S_j} = A^*A, \text{ where } A = d\psi(D^2)\big|_{S_j}.$$
Now A is a trace-class operator, so we may write in any orthonormal basis (e_i) for $L^2(S)$
$$\text{Tr}(A^*A) = \sum_i \langle A^*Ae_i, e_i \rangle = \sum \|Ae_i\|^2 \geq 0.$$
Therefore $(\mu_j - \beta_j) - (\mu_{j-1} - \beta_{j-1}) + (\mu_{j-2} - \beta_{j-2}) - \cdots \geq 0$, and the result follows.
□

Morse functions

From now on, we will consider only the case of the de Rham complex.

DEFINITION 14.4 A smooth function $h : M \to \mathbb{R}$ is called a *Morse function* on M if at its critical points (that is, the points where the first derivatives ∇h vanish) the Hessian[1] H_h (the matrix of second derivatives) is non-singular.

Clearly the critical points of a Morse function are isolated, so there are only finitely many of them. Each critical point has an *index*, defined to be the number of negative eigenvalues of the Hessian at that point.

Let d_s be Witten's perturbed exterior derivative associated to the function h, as in 9.14. Let d_s^* be its adjoint, and $D_s = d_s + d_s^*$ the perturbed de Rham operator. We will look at the asymptotics of the perturbed de Rham complex as $s \to \infty$. Eventually we will need to choose a special metric on M that is nicely related to the Morse function h, but we can do the first part of the calculation without making this special choice.

We will need to know that the basic elliptic theory of Chapter 5, and the results on finite propagation speed such as 7.23, extend without change to the operator D_s. Indeed, the operator D_s differs from the standard Dirac operator D only by a

---
[1] See 9.16.

zero order perturbation, so it belongs to the class of self-adjoint generalized Dirac operators which was already considered in Chapter 5; the proof of finite propagation speed goes over verbatim. As an example, let us verify the Garding inequality for D_s. By Lemma 9.17,

$$D_s^2 = d^2 + s\mathsf{H}_h + s^2|dh|^2 = D^2 + L$$

where L is an operator of order zero. Therefore

$$\|D_s\omega\|^2 = \|D\omega\|^2 + \langle L\omega, \omega\rangle \geqslant \|D\omega\|^2 - C_1\|\omega\|^2$$

for some constant C_1. Hence

$$(1 + C_1)(\|D_s\omega\|^2 + \|\omega\|^2) \geqslant \|D\omega\|^2 + \|\omega\|^2 \geqslant \frac{1}{C_2}\|\omega\|_1^2$$

by the usual Garding inequality (5.14); the Garding inequality for D_s follows.

REMARK 14.5 Notice that the norm of L is of order s^2, so the constant appearing in the Garding inequality is bounded by a polynomial in s. The same is true of the constants appearing in the elliptic estimates.

We begin our asymptotic calculation of Witten's complex by fixing a number $\rho > 0$ and choosing a positive even function $\varphi \in \mathcal{S}(\mathbb{R})$ with $\varphi(0) = 1$ and such that the Fourier transform $\hat{\varphi}$ is supported within the interval $[-\rho, \rho]$. According to 14.3, the Betti numbers of M satisfy the Morse inequalities relative to the numbers $\mu_j = \mathrm{Tr}\left(\varphi(D_s)\big|_{\Lambda^j T^*M}\right)$. (Note that since φ is even, $\varphi(D_s)$ can in fact be written as a function of D_s^2, so (14.3) is applicable.) We investigate the asymptotics of $\varphi(D_s)$ as $s \to \infty$. First, we work on the complement of a neighbourhood of the set of critical points of h. Let us denote this set of critical points by $\mathrm{Crit}(h)$.

LEMMA 14.6 *On the complement of a 2ρ-neighbourhood of* $\mathrm{Crit}(h)$, *the smoothing kernel of $\varphi(D_s)$ tends uniformly to zero as $s \to \infty$.*

PROOF Since M is a compact manifold, there is a constant C such that $|\nabla h(x)| \geqslant C > 0$ for all x in the complement of a ρ-neighbourhood of $\mathrm{Crit}(h)$. Now by the formula

$$D_s^2 = d^2 + s\mathsf{H}_h + s^2|dh|^2$$

and lemma 9.17 we find that for s large

$$\langle D_s^2 \omega, \omega \rangle \geq \tfrac{1}{2} C^2 s^2 \|\omega\|^2 \qquad (14.7)$$

provided that ω is supported in the complement of such a neighbourhood. Now let $\mathfrak{H}$ denote the Hilbert space of L^2 differential forms on M that vanish on a ρ-neighbourhood of $\text{Crit}(h)$. Formula 14.7 shows that D_s^2 is a positive formally self-adjoint operator on $\mathfrak{H}$. It therefore has a self-adjoint extension A on $\mathfrak{H}$ satisfying the same positivity condition, by Friedrichs' extension theorem [29]. Now we will show that if ω is supported in the complement of a 2ρ-neighbourhood of $\text{Crit}(h)$, then

$$\varphi(D_s)\omega = \varphi(\sqrt{A})\omega.$$

To do this we use unit propagation speed (7.20) for the operator D_s. Consider the time-dependent differential form

$$\omega_t = \cos(tD_s)\omega = \frac{1}{2}(e^{itD_s} + e^{-itD_s})\omega.$$

Clearly ω_t is a solution to the partial differential equation

$$\frac{\partial^2 \omega_t}{\partial t^2} + D_s^2 \omega_t = 0$$

with initial conditions $\omega_0 = \omega$, $\dot{\omega}_0 = 0$; in fact it is the unique solution, as one can easily check by verifying that the "energy"

$$\left\| \frac{\partial \omega_t}{\partial t} \right\|^2 + \langle D_s^2 \omega_t, \omega_t \rangle$$

is conserved (compare the proof of 7.4). But by the unit propagation speed property, ω_t is supported in the complement of a ρ-neighbourhood of $\text{Crit}(h)$ for $|t| < \rho$, and therefore $D_s^2 \omega_t = A \omega_t$. Thus ω_t for $|t| < \rho$ is also the unique solution to the equation

$$\frac{\partial^2 \omega_t}{\partial t^2} + A \omega_t = 0$$

with the same initial conditions, so we may write $\omega_t = \cos(t\sqrt{A})\omega$.

Now $\hat{\varphi}$ has support in $[-\rho, \rho]$ and moreover is an even function (since φ is).

Therefore

$$\begin{aligned}\varphi(D_s)\omega &= \frac{1}{2\pi}\int_{-\rho}^{\rho}(e^{itD_s}\omega)\hat{\varphi}(t)\,dt\\ &= \frac{1}{\pi}\int_{0}^{\rho}\hat{\varphi}(t)\cos(tD_s)\omega\,dt\\ &= \frac{1}{\pi}\int_{0}^{\rho}\hat{\varphi}(t)\omega_t\,dt\\ &= \frac{1}{\pi}\int_{0}^{\rho}\hat{\varphi}(t)\cos(t\sqrt{A})\omega\,dt\\ &= \ldots = \varphi(\sqrt{A})\omega.\end{aligned}$$

This proves our claim. But now notice that $\sqrt{A}$ is a positive operator, bounded below by $\frac{1}{2}Cs$. It follows from the spectral theorem, then, that the L^2 operator norm of $\varphi(\sqrt{A})$ is bounded above by

$$c(s) = \sup\{|\varphi(\lambda)| : \lambda \geqslant \frac{1}{2}Cs\}.$$

As $s \to \infty$, this quantity tends to zero with rapid decay. So we deduce that if ω is supported in the complement of a 2ρ-neighbourhood of Crit(h),

$$\|\varphi(D_s)\omega\| \leqslant c(s)\|\omega\| \tag{14.8}$$

with $c(s) \to 0$ rapidly as $s \to \infty$.

This is nearly what we want. In fact, if we can show that there is a $c_1(s)$, tending to zero as $s \to \infty$, with

$$\|\varphi(D_s)\omega\|_{L^\infty} \leqslant c_1(s)\|\omega\|_{L^1} \tag{14.9}$$

(under the same condition on Supp(ω)), we will be done, since for any integral operator with continuous kernel the supremum of the kernel can be estimated by the norm of the operator as a map from L^1 to L^∞; this is simply a rephrasing of the fact that $(L^1)^* = L^\infty$.

To get the improved estimate 14.9 from 14.8 we rely on the familiar technique of Sobolev embedding. The key point is this: for any k, the operator $(1+D_s^2)^{-1}$ bounded as an operator from W^k to W^{k+2}, with norm bounded by a polynomial in s. This follows from the elliptic estimates for D_s. Now by Sobolev embedding (5.7) $W^p \subset L^\infty$ for $p > \frac{n}{2}$, and therefore $(1+D_s^2)^{-k}$ is bounded from L^2 to L^∞ for $k >$

the bound being polynomial in s. By duality and self-adjointness, $(1+D_s^2)^{-k}$ is also bounded (polynomially in s) as an operator from L^1 to L^2.

We deduce that the norm of $\varphi(D_s)$ acting as an operator from L^1 to L^∞ is bounded by a polynomial in s times the norm of $(1+D_s^2)^{2k}\varphi(D_s)$ acting as an operator on L^2. But this operator is just $\tilde{\varphi}(D_s)$, where $\tilde{\varphi}(\lambda) = (1+\lambda^2)^{2k}\varphi(\lambda)$; the function $\tilde{\varphi}$ satisfies the same conditions as φ, so

$$\|\tilde{\varphi}(D_s)\omega\| \leqslant \tilde{c}(s)\|\omega\|$$

provided that ω satisfies the support condition, with $\tilde{c}(s)$ of rapid decay in s. Therefore

$$\|\varphi(D_s)\omega\|_{L^\infty} \leqslant c_1(s)\|\omega\|_{L^1}$$

with $c_1(s) = \tilde{c}(s) \times $ (polynomial in s), which tends to zero as $s \to \infty$. □

From Lemma (14.6) it follows that as $s \to \infty$, the trace of $\varphi(D_s)$ is given by a sum of contributions from the critical points of h. The reader should notice the similarity with the Lefschetz theorem (Chapter 8). We will now evaluate the contributions from the critical points.

The contribution from the critical points

It is convenient to make a special choice of metric on our manifold M. This choice of metric uses the *Morse lemma*.

LEMMA 14.10 *There are local co-ordinates (x_j) centered at each critical point of h with the property that in terms of these local co-ordinates h is a diagonal quadratic form*

$$h(x) = \frac{1}{2}\sum \lambda_j (x^j)^2.$$

Of course the number of negative λ's is just the index of the critical point.

We will not go through the proof of the Morse lemma here. A proof may be found in Milnor [54].

Now we choose our special metric g on M as follows; g is defined to be flat Euclidean ($g_{ij} = \delta_{ij}$) in Morse co-ordinates near each critical point, and is patched up away from

Crit(h) using a partition of unity. We choose ρ so small that g is flat Euclidean at least to distance 4ρ from each critical point.

We calculated in chapter 9 that, when h is a quadratic form on flat Euclidean space, the operator D_s^2 is equal to

$$L_s = \sum_j \left\{ -\left(\frac{\partial}{\partial x^j}\right)^2 + s^2 \lambda_j^2 (x^j)^2 + s\lambda_j Z_j \right\}$$

where $Z_j = [dx^j \lrcorner \cdot, dx^j \wedge \cdot]$. Moreover we recall from proposition 9.18 that the spectrum of L_s can be described explicitly: L_s is an essentially self-adjoint operator, with discrete spectrum. The eigenvalues of L_s are the numbers

$$s \sum_j (|\lambda_j|(1 + 2p_j) + \lambda_j q_j)$$

where $p_j = 0, 1, 2, \ldots$ and $q_j = \pm 1$. If we consider the action of L_s on k-forms, the spectrum is as above with the additional restriction that exactly k of the q_j's are equal to $+1$.

LEMMA 14.11 *Suppose that precisely m of the λ_j's are negative. Then*

$$\lim_{s \to \infty} \operatorname{Tr}\left(\varphi(\sqrt{L_s})\big|_{\Lambda^k}\right) = \begin{cases} 0 & (k \neq m) \\ 1 & (k = m). \end{cases}$$

Moreover, the same limit holds good for $\operatorname{Tr}\left(B\varphi(\sqrt{L_s})\big|_{\Lambda^k}\right)$ *where B is the operator of multiplication on $\mathbb{R}^n$ by any $\beta \in C_c^\infty(\mathbb{R}^n)$ with $\beta(0) = 1$.*

PROOF By 8.7 and 9.18,

$$\operatorname{Tr}\left(\varphi(\sqrt{L_s})\big|_{\Lambda^k}\right) = \sum_{p_j, q_j} \varphi\left(\sqrt{s \sum_j (|\lambda_j|(1 + 2p_j) + \lambda_j q_j)}\right)$$

where the summation is over $p_j = 0, 1, 2, \ldots$ and $q_j = \pm 1$ and exactly k of the q_j equal $+1$. If $k \neq m$ then all the eigenvalues of L_s are of order s. Since φ is rapidly decreasing, it is easy to check that the sum tends to 0 as $s \to \infty$. On the other hand if $k = m$ then precisely one eigenvalue equals 0 and the others are of order s. The 0-eigenvalue contributes 1 to the sum and the sum of the remaining terms tends to 0, for the same reason as before.

In the case of the more general trace $\text{Tr}\left(B\varphi(\sqrt{L_s})\big|_{\Lambda^k}\right)$, let $e(p_j, q_j)$ denote the normalized eigenform of L_s corresponding to (p_j, q_j). Then 8.7 gives

$$\text{Tr}\left(B\varphi(\sqrt{L_s})\big|_{\Lambda^k}\right) = \sum_{p_j, q_j} \varphi\left(\sqrt{s\sum_j(|\lambda_j|(1 + 2p_j) + \lambda_j q_j)}\right) \langle Be(p_j, q_j), e(p_j, q_j)\rangle.$$

For the same reason as before, only the zero eigenvalue makes a contribution to this sum that does not vanish as $s \to \infty$. The corresponding eigenform e_0 is just the ground state eigenfunction of the harmonic oscillator multiplied by a certain constant differential form; namely, by $dx^1 \wedge \ldots \wedge dx^m$ if we assume that the first m of the λ_j's are negative and the rest are positive. That is, by 9.7, the eigenform e_0 is given explicitly by

$$e_0 = (s^{n/2}\pi^{-n/4}\prod_j \lambda_j^{\frac{1}{2}})\exp(-s\sum_j \lambda_j(x^j)^2/2)\, dx^1 \wedge \ldots \wedge dx^m.$$

It is now easy to check that as $s \to \infty$, $\langle Be_0, e_0\rangle \to 1$; so the stated result follows.
]

We can now state and prove Morse's theorem.

THEOREM 14.12 *Let h be a Morse function on the compact manifold M. Let β_j denote the j'th Betti number of M and let ν_j denote the number of critical points of of index j. Then*

$$\beta_0 \leq \nu_0$$
$$\beta_1 - \beta_0 \leq \nu_1 - \nu_0$$
$$\beta_2 - \beta_1 + \beta_0 \leq \nu_2 - \nu_1 + \nu_0$$
$$\vdots$$
$$\sum(-1)^j\beta_j = \sum(-1)^j\nu_j.$$

PROOF Choose a metric on M, euclidean near the critical points, and a cut-off function φ as above. By 14.3, the Betti numbers β_j^s of Witten's perturbed de Rham complex satisfy the Morse inequalities with respect to the numbers $\mu_j^s = \left(\varphi(D_s)\big|_{\Lambda^j}\right)$.

But the perturbed de Rham complex is conjugate to the unperturbed one; the two complexes therefore have isomorphic homology groups. In particular $\beta_j^s = \beta_j$ for all j. The proof will therefore be completed if we can show that $\mu_j^s \to \nu_j$ as $s \to \infty$.

By 8.12, the trace of $\varphi(D_s)|_{\Lambda^j}$ is obtained by integration of the local trace of the smoothing kernel of this operator over the diagonal. By (14.6) this local trace tends uniformly to zero except on a 2ρ-neighbourhood of each critical point. So the limit as $s \to \infty$ of $\varphi(D_s)|_{\Lambda^j}$ is a sum of contributions from the critical points. The contribution from a critical point can be written $\lim_{s\to\infty} \text{Tr}\left(B\varphi(D_s)\big|_{\Lambda^j}\right)$, where B is the multiplication operator by a smooth function on M equal to 1 on a 2ρ-neighbourhood of the critical point and supported in a 3ρ-neighbourhood of the critical point.

Now take Morse co-ordinates around the critical point. These enable us to identify forms supported on a 4ρ-neighbourhood of the critical point in M with forms supported on a 4ρ-neighbourhood of the origin in $\mathbb{R}^n$. Under this identification D_s^2 corresponds to L_s.

Now a unit propagation speed argument exactly analogous to that given in 14.6 shows that
$$\varphi(D_s)\alpha = \varphi\left(\sqrt{L_s}\right)\alpha$$
provided that α is supported in a 3ρ-neighbourhood of the critical point. Hence
$$\text{Tr}\left(B\varphi(D_s)\big|_{\Lambda^j}\right) = \text{Tr}\left(B\varphi(\sqrt{L_s})\big|_{\Lambda^j}\right).$$
But by Lemma (14.11), as $s \to \infty$, $\text{Tr}\left(B\varphi(\sqrt{L_s})\big|_{\Lambda^j}\right)$ tends to 1 if the critical point has index j, and otherwise to 0. The result follows. $\square$

Notes

For an inspiring overview of Morse theory, read the lectures of Bott [14].

CHAPTER 15

Atiyah's Γ-index theorem

In 11.15 we showed that the index is multiplicative under coverings: if D is a Dirac operator on the compact manifold M, and $\tilde{D}$ the lifted operator on a k-fold covering $\tilde{M}$, then $\mathrm{Ind}(\tilde{D}) = k.\mathrm{Ind}(D)$. We may express this by saying that the amount of index per unit area is the same on M as on $\tilde{M}$. This, however, immediately suggests a possible generalization to infinite coverings of M, provided that we can make sense of the concept "average amount of index per unit area". The generalization is Atiyah's Γ-index theorem.

An algebra of smoothing operators

Throughout this chapter, the following notation will be fixed. M denotes a compact oriented Riemannian manifold, and S is a Clifford bundle over M with Dirac operator D. $\tilde{M}$ denotes a Galois covering of M with Galois group Γ; this means that Γ is a homomorphic image of $\pi_1(M)$ and $\tilde{M}$ is the natural cover of M with fiber Γ; Γ acts discontinuously on $\tilde{M}$ by deck transformations and $\tilde{M}/\Gamma = M$. Let $\tilde{S}$ and $\tilde{D}$ denote the natural lifts of S and D to $\tilde{M}$, and $\pi\colon \tilde{M} \to M$ be the covering map. By 9.20, the operator $\tilde{D}$ is essentially self adjoint on $L^2(\tilde{S})$ and operators $f(\tilde{D})$ can be defined for every $f \in C_0(\mathbb{R})$.

In our discussion of the ordinary index theorem we saw the key rôle played by the algebra of smoothing operators. On the non-compact manifold $\tilde{M}$, a central idea is to introduce an algebra $\mathcal{A}$ of smoothing operators on $\tilde{M}$ that reflects the extra structure of the Γ-action.

DEFINITION 15.1 With notation as above, let $\mathcal{A}$ be the set of bounded operators A on $L^2(\tilde{S})$ satisfying the following conditions:

(i) A is Γ-equivariant; that is, for all $s \in L^2(\tilde{S}), A(\gamma s) = \gamma(As)$ where by definition
$$\gamma s(p) = s(p\gamma^{-1});$$

(ii) A is represented by a smoothing kernel $k(p,q)$ so that

$$As(p) = \int k(p,q)s(q)\,\text{vol}(q);$$

(iii) There is an absolute constant C such that

$$\int |k(p,q)|^2 \,\text{vol}(q) < C, \quad \int |k(p,q)|^2 \,\text{vol}(p) < C.$$

We must be careful what we mean by "smoothing kernel" on the non-compact manifold $\tilde{M}$, since differentiation under the integral sign is not automatically legitimate. We will therefore assume (as part of condition (ii)) that

$$m \mapsto k(p,.) \text{ and } q \mapsto k(.,q)$$

are C^∞ maps of $\tilde{M}$ to the Hilbert space $L^2(\tilde{S})$.

LEMMA 15.2 *The set of operators $\mathcal{A}$ forms an algebra.*

PROOF The only thing that is not obvious is that $\mathcal{A}$ is closed under multiplication. If $A_1, A_2 \in \mathcal{A}$ are represented by the smoothing kernels k_1, k_2 then $A_1.A_2$ is represented by the smoothing kernel

$$(p,r) \mapsto \int k_1(p,q)k_2(q,r)\,\text{vol}(q).$$

Let A be a bounded operator on the Hilbert space $L^2(\tilde{S})$ which is represented by a smoothing kernel k. By the Riesz representation theorem for functionals on Hilbert space, the quantity

$$\int |k(p,q)|^2 \,\text{vol}(q)$$

is the square of the norm of the linear functional $s \mapsto As(p)$ on $L^2(\tilde{S})$. Therefore there is a constant C such that $\int |k(p,q)|^2 \,\text{vol}(q) < C$ if and only if A maps $L^2(\tilde{S})$ continuously to the space $CB(\tilde{S})$ of bounded continuous sections of $\tilde{S}$. Similarly there is a constant C such that $\int |k(p,q)|^2 \,\text{vol}(p) < C$ if and only if A^* maps $L^2(\tilde{S})$ continuously to $CB(\tilde{S})$. Now the desired result is clear; for if A_1, A_2 and their adjoints map $L^2(\tilde{S})$ continuously to $L^2(\tilde{S})$ and to $CB(\tilde{S})$, then so do $A_1.A_2$ and $(A_1.A_2)^* = A_2^*.A_1^*$. □

DEFINITION 15.3 *The space $CB^r(\tilde{S})$ is defined to be the space of sections s of $\tilde{S}$ which are r times continuously differentiable with bounded derivatives.*

This definition needs some explanation. The r'th derivative $\nabla^r(s)$ of a section s of $\tilde{S}$ is defined in terms of the connection on $\tilde{S}$ as a tensor of type $\binom{0}{r}$ with values in , that is, a section of the bundle $\bigotimes^r(T^*\tilde{M}) \otimes \tilde{S}$. This tensor bundle has a natural metric, and we require that the derivative be uniformly bounded in terms of this metric. Clearly $CB^r(\tilde{S})$ is a Banach space under the natural supremum norm of the first r derivatives.

We now prove a non-compact Sobolev embedding theorem.

PROPOSITION 15.4 *Let $n = \dim(M) = \dim(\tilde{M})$. For any integer $p > \frac{n}{2}$ and any ≥ 0 there is a constant C such that*

$$\|s\|_{CB^r} \leq C\big(\|s\| + \|\tilde{D}s\| + \cdots + \|\tilde{D}^{p+r}s\|\big)$$

for all s in the domain of $\tilde{D}^{p+r}$, the norms on the right-hand side being L^2 norms.

PROOF To show that s is of class CB^r, it is enough to show that it is uniformly of class C^r locally. Choose $m \in \tilde{M}$ and pick a bump function φ_m on $\tilde{M}$, equal to one in a neighbourhood of m and supported within a fundamental domain for the action of Notice that we can choose such bump functions φ_m in such a way that their first $+ r)$ derivatives are bounded uniformly in m. Now

$$\|\varphi_m s\| + \cdots + \|\tilde{D}^{p+r}(\varphi_m s)\| \leq C_m\big(\|s\| + \cdots + \|\tilde{D}^{p+r}s\|\big)$$

Here the constant C_m depends on the first $(p+r)$ derivatives of φ_m and may therefore be taken to be bounded uniformly in m. But now $\varphi_m s$ may be identified by means of the covering map π with a section $\pi_*(\varphi_m s)$ of S over M. By the elliptic estimates and Sobolev embedding theorem on M, then

$$\|\pi_*(\varphi_m s)\|_{C^r(S)} \leq (\text{const.})\big(\|\varphi_m s\| + \cdots + \|\tilde{D}^{p+r}(\varphi_m s)\|\big).$$

The result now follows, as $\varphi_m \equiv 1$ near m. □

LEMMA 15.5 *Let A be a bounded, self-adjoint, equivariant operator on $L^2(\tilde{S})$, and suppose that A maps L^2 boundedly to CB^r for each r. Then A^2 belongs to the algebra $\mathcal{A}$.*

195

PROOF Clearly A^2 is equivariant. For the remainder of the proof, assume fo simplicity that S is a trivial bundle; the general case is a bit more complicate notationally. Choose a point $p \in M$. The functional on $L^2(S)$ given by $S \mapsto As(p$ is continuous linear, so it is represented by a vector $v_p \in L^2(S)$, in the sense tha $As(p) = \langle s, v_p \rangle$. The norms of the vectors v_p are bounded uniformly in p; this because A maps L^2 to the *bounded* continuous functions.

We claim that for any $r \geqslant 0$, the map $p \mapsto v_p$ is a C^r-differentiable map of M t the Banach space $L^2(S)$. We will write out the details only for the case $r = 0$. The for any $s \in L^2(S)$, one has an estimate on the first derivative of As in terms of th L^2-norm of s. By the mean-value theorem, therefore,

$$|\langle s, v_p - v_q \rangle| = |As(p) - As(q)| \leqslant C \|s\|_{L^2} \, d(p, q),$$

for some constant C. Therefore, $\|v_p - v_q\| \leqslant C \, d(p, q)$, which proves (with room t spare!) that $m \mapsto v_m$ is a continuous map of M to L^2.

Now we write

$$As(p) = \int s(q) \bar{v}_p(q) \, \text{vol}(q).$$

Since A is self-adjoint, $\langle As_1, s_2 \rangle = \langle s_1, As_2 \rangle$. Therefore

$$\int s_1(q) \bar{s}_2(p) \bar{v}_p(q) \, \text{vol}(p) \, \text{vol}(q) = \int s_1(q) \bar{s}_2(q) v_p(q) \, \text{vol}(p) \, \text{vol}(q).$$

This gives $v_p(q) = \bar{v}_q(p)$ so we may write

$$As(p) = \int s(q) v_q(p) \, \text{vol}(q).$$

So

$$A^2 s(p) = \int k(p, q) s(q) \, \text{vol}(q),$$

where

$$k(p, q) = A v_q(p).$$

The kernel k is of class C^r, since $q \mapsto v_q$ is a C^r map of M to $L^2(S)$, so $q \mapsto A v_q$ a C^r map of M to $CB^r(S)$. Moreover, the functions $k(\cdot, q) = A v_q$ form a bound subset of $L^2(S)$, and they are equal by self-adjointness to the functions $k(q, \cdot)$. Th completes the proof. □

PROPOSITION 15.6 For any rapidly decreasing function f on $\mathbb{R}$, the operator $f(\tilde{D})$ belongs to the algebra $\mathcal{A}$.

PROOF We may assume that f is non-negative. Then $f^{1/2}$ is rapidly decreasing too, so by proposition 15.4 $f^{1/2}(D)$ maps $L^2(S)$ boundedly to $CB^r(S)$ for all s. Also $f^{1/2}(D)$ is bounded, self-adjoint, and equivariant. Hence

$$f(D) = (f^{1/2}(D))^2 \in \mathcal{A}$$

by 15.5. □

DEFINITION 15.7 The functional $\tau\colon \mathcal{A} \to \mathbb{C}$ is defined as follows: if $A \in \mathcal{A}$, let k be its kernel and choose any fundamental domain F for the Γ-action on $\tilde{M}$; then

$$\tau(A) = \int_F \operatorname{tr} k(p,p).\operatorname{vol}(p).$$

Notice that since A is equivariant, its kernel k is equivariant in the sense that $(p\gamma, q\gamma) = k(p,q)$. Therefore the definition of τ does not depend on the choice of fundamental domain. The functional τ on $\mathcal{A}$ will play the rôle of the trace on the algebra of smoothing operators on a compact manifold. The next result is analogous to the important property 8.8:

PROPOSITION 15.8 For $A_1, A_2 \in \mathcal{A}$,

$$\tau(A_1 A_2) = \tau(A_2 A_1).$$

PROOF Let A_1 and A_2 be represented by kernels k_1 and k_2. Then $A_1 A_2 - A_2 A_1$ is presented by the kernel

$$(p,r) \to \int_{\tilde{M}} \Big(k_1(p,q)k_2(q,r) - k_2(p,q)k_1(q,r)\Big) \operatorname{vol}(q).$$

Therefore, if F is a fundamental domain,

$$\tau(A_1 A_2 - A_2 A_1) = \iint_{F \times \tilde{M}} \operatorname{tr}\Big(k_1(p,q)k_2(q,p) - k_2(p,q)k_1(q,p)\Big) \operatorname{vol}(q) \operatorname{vol}(p)$$

Because of the estimate (15.1)(iii) this double integral converges absolutely. So we may decompose $\tilde{M} = \cup_{\gamma \in \Gamma} F\gamma$, and write

$$\tau(A_1 A_2 - A_2 A_1) = \sum_{\gamma \in \Gamma} \varphi(\gamma),$$

where $\varphi(\gamma)$ equals

$$\iint_{F\times F} \mathrm{tr}\,\big(k_1(p,q)k_2(q\gamma,p) - k_2(p,q\gamma)k_1(q\gamma,p)\big)\,\mathrm{vol}(q)\,\mathrm{vol}(p).$$

But, by equivariance of k_1 and k_2, this equals

$$\iint_{F\times F} \mathrm{tr}\,\big(k_1(p\gamma^{-1},q)k_2(q,p\gamma^{-1}) - k_2(p\gamma^{-1},q)k_1(q,p\gamma^{-1})\big)\,\mathrm{vol}(q)\,\mathrm{vol}(p)$$

which is $-\varphi(\gamma^{-1})$. So

$$\sum_{\gamma\in\Gamma}\varphi(\gamma) = -\sum_{\gamma\in\Gamma}\varphi(\gamma^{-1}),$$

hence is equal to 0. □

Renormalized dimensions and the index theorem

DEFINITION 15.9 If $\mathcal{H}$ is a subspace of $L^2(\tilde{S})$ with the property that the orthogonal projection operator P from $L^2(\tilde{S})$ onto $\mathcal{H}$ belongs to $\mathcal{A}$, then we define

$$\dim_\Gamma(\mathcal{H}) = \tau(P).$$

Of course, this definition is motivated by the fact that the trace of a projection operator is the dimension of its range. We will see that $\dim_\Gamma$ conforms to our intuitive idea of measuring the "average amount of dimension per unit area". In particular

LEMMA 15.10 *Under the hypotheses of 15.9, if* $\dim_\Gamma(\mathcal{H}) > 0$ *and* Γ *is infinite, then* $\mathcal{H}$ *is infinite-dimensional (in the usual sense).*

PROOF Suppose to the contrary that $\mathcal{H}$ is finite-dimensional, and let $s_1,\ldots,s_j$ be an orthonormal basis for it. Then the operator P has smoothing kernel

$$k(p,q) = \sum_i s_i(p)\otimes s_i(q),$$

(where we have identified S^* with S by means of the metric) and so

$$\mathrm{tr}\,k(p,p) = \sum_i |s_i(p)|^2$$

is an integrable function of p. Let F denote a fundamental domain. Then

$$\sum_{\gamma\in\Gamma}\int_{F\gamma} |\mathrm{tr}\,k(p,p)|\,\mathrm{vol}(p) < \infty,$$

and since the integral is independent of γ it must be zero. □

Now suppose that D (and hence $\tilde{D}$) is a graded Dirac operator. By 15.4 and 15.5, the orthogonal projection P onto the kernel of $\tilde{D}$ belongs to the algebra $\mathcal{A}$. We can therefore define

$$\operatorname{Ind}_\Gamma(\tilde{D}) = \dim_\Gamma(\ker \tilde{D}_+) - \dim_\Gamma(\ker \tilde{D}_-).$$

Let ε be the grading operator. By analogy with 11.1, for $A \in \mathcal{A}$ we define $\tau_s(A) = \tau(\varepsilon A)$. The following is the analogue of the McKean-Singer formula.

PROPOSITION 15.11 *For any $t > 0$,*

$$\tau_s(e^{-t\tilde{D}^2}) = \operatorname{Ind}_\Gamma(\tilde{D}).$$

PROOF Because τ has the basic 'trace property' (15.8), the proof that $\tau_s(e^{-t\tilde{D}^2})$ independent of t is exactly the same as that given in 11.9; in fact, $\tau_s(f(D^2))$, is independent of the choice of rapidly decreasing function f with $f(0) = 1$. To complete the proof we need an analogue of Lemma 10.5, and here we must proceed differently. What we need to show is that as $t \to \infty$, the smoothing kernel of $e^{-t\tilde{D}^2}$ tends to the smoothing kernel of P uniformly on compact subsets of $\tilde{M} \times \tilde{M}$.

We claim first that as $t \to \infty$, $e^{-t\tilde{D}^2} \to P$ in the strong operator topology on $L^2(\tilde{S})$, which means that

$$e^{-t\tilde{D}^2} x \to Px \quad \text{for all } x \in L^2. \tag{15.12}$$

This is a consequence of the spectral theorem for self-adjoint operators [29], but an elementary argument may be given as follows. Since $\tilde{D}$ is self-adjoint, the orthogonal complement of the kernel of $\tilde{D}$ is the closure of its range. This means that it suffices to verify 15.12 in two special cases: when x belongs to the kernel, and when x belongs to the range. In the first case $e^{-t\tilde{D}^2} x = x = Px$ for all t; in the second case, let $x = \tilde{D}y$ and note that then

$$\|e^{-t\tilde{D}^2} x\| = \|\tilde{D} e^{-t\tilde{D}^2} y\| \to 0$$

since $\sup_\lambda |\lambda e^{-t\lambda^2}| = O(t^{1/2})$. The result follows.

Now consider the smoothing kernels k_t of $e^{-t\tilde{D}^2}$ as elements of the Fréchet space $\mathcal{E}(S \boxtimes S^*)$. From the proof of proposition 15.6, as $t \to \infty$, the kernels k_t form a bounded subset of this Fréchet space. But in this Fréchet space, bounded subsets are relatively compact; so we deduce that given any sequence $t_j \to \infty$, there is a subsequence of the t_j's such that the kernel of $e^{-t\tilde{D}^2}$ tends to a limit uniformly on compact subsets as $t \to \infty$ through this subsequence. By weak convergence, this limit must be the kernel of P. Finally, to complete the proof, we apply a lemma of general topology: if $t \to x(t)$ is a curve in a metric space X, and there is $x_0 \in X$ such that for each sequence $t_j \to \infty$ there is a subsequence t_{j_k} such that $x(t_{j_k}) \to x_0$ as $k \to \infty$, then $x(t) \to x_0$ as $t \to \infty$. □

THEOREM 15.13 (ATIYAH'S Γ-INDEX THEOREM) *Under the hypotheses of this chapter,*

$$\mathrm{Ind}_\Gamma(\tilde{D}) = \mathrm{Ind}(D).$$

PROOF By (15.11),

$$\mathrm{Ind}_\Gamma(\tilde{D}) = \tau_s(e^{-t\tilde{D}^2}),$$

for any $t > 0$. Now the asymptotic expansion 7.15 still applies to $e^{-t\tilde{D}^2}$; the estimate (15.4) plays the rôle of the elliptic estimate in its proof. So as in 11.14 we get the formula

$$\mathrm{Ind}_\Gamma(\tilde{D}) = \frac{1}{(4\pi)^{n/2}} \int_F \mathrm{tr}_s \, \tilde{\Theta}_{n/2}$$

where $\tilde{\Theta}$ is the asymptotic-expansion coefficient for $e^{-t\tilde{D}^2}$. But the asymptotic expansion coefficient is simply a local algebraic expression in the metrics and connection coefficients and their derivatives, so

$$\tilde{\Theta}_{n/2} = \pi^* \Theta_{n/2}$$

where Θ is the corresponding asymptotic-expansion coefficient for D on the compact manifold M. Therefore

$$\frac{1}{(4\pi)^{n/2}} \int_F \mathrm{tr}_s \, \tilde{\Theta}_{n/2} = \frac{1}{(4\pi)^{n/2}} \int_M \mathrm{tr}_s \, \Theta_{n/2} = \mathrm{Ind}(D)$$

by 11.14. □

EXAMPLE 15.14 Let M be a Riemann surface of genus $g \geqslant 2$, equipped with its Poincaré metric (of constant curvature -1). Then $\tilde{M}$ is 2-dimensional hyperbolic space, that is the unit disc with its Poincaré metric. Let D be the de Rham operator on M, equipped with the grading it inherits from the de Rham complex, so that $\text{Ind}(D) =$ Euler characteristic of $M = 2 - 2g$. By the Γ-index theorem, $\text{Ind}(\tilde{D}) = 2 - 2g < 0$; which implies by 15.10 that the space of square-integrable harmonic 1-forms on the disc is infinite-dimensional. Of course we knew this already, but in some sense this shows us what topology the space of L^2 harmonic 1-forms is detecting. For more on this, see [26].

EXAMPLE 15.15 An unsolved conjecture in geometry, apparently due to Hopf, is that if M is a compact $2m$-dimensional Riemannian manifold of negative sectional curvature, then the sign of the Euler characteristic of M is $(-1)^m$. Singer suggested an approach to this problem by way of the L^2 Gauss-Bonnet theorem; show that the space of L^2 harmonic forms on the universal cover $\tilde{M}$ vanishes except in the middle dimension. Some progress has been made in this direction, see [28, 37].

EXAMPLE 15.16 In a major application of the Γ-index theorem, Atiyah and Schmid constructed certain representations (the so-called "discrete series") of Lie groups as spaces of L^2 holomorphic sections of certain vector bundles. The Γ-index theorem was used to show that these spaces of sections are non-zero. See Schmid [66].

Notes

The L^2 index theorem is due to Atiyah [1]. Several of its techniques — the introduction of operator algebras, of traces and of generalized "dimension functions" — have served as paradigms for other index theorems on non-compact manifolds, such as the foliation index theorem of Connes [20, 22] or the exhaustion index theorem [61, 62].

The theory of 'L^2 homological algebra' inspired by the L^2-index theorem is quite active at present. For a survey see Lück [50].

References

1. M.F. Atiyah. Elliptic operators, discrete groups and von Neumann algebras. *Astérisque*, 32:43–72, 1976.
2. M.F. Atiyah and R. Bott. A Lefschetz fixed-point formula for elliptic complexes I. *Annals of Mathematics*, 86:374–407, 1967.
3. M.F. Atiyah, R. Bott, and V.K. Patodi. On the heat equation and the index theorem. *Inventiones Mathematicae*, 19:279–330, 1973.
4. M.F. Atiyah, R. Bott, and A. Shapiro. Clifford modules. *Topology*, 3 (supplement 1):3–38, 1964.
5. M.F. Atiyah, V.K. Patodi, and I.M. Singer. Spectral asymmetry and Riemannian geometry I. *Mathematical Proceedings of the Cambridge Philosophical Society*, 77:43–69, 1975.
6. M.F. Atiyah, V.K. Patodi, and I.M. Singer. Spectral asymmetry and Riemannian geometry II. *Mathematical Proceedings of the Cambridge Philosophical Society*, 78:405–432, 1975.
7. M.F. Atiyah, V.K. Patodi, and I.M. Singer. Spectral asymmetry and Riemannian geometry III. *Mathematical Proceedings of the Cambridge Philosophical Society*, 79:71–99, 1976.
8. M.F. Atiyah and G.B. Segal. The index of elliptic operators II. *Annals of Mathematics*, 87:531–545, 1968.
9. M.F. Atiyah and I.M. Singer. The index of elliptic operators I. *Annals of Mathematics*, 87:484–530, 1968.
10. M.F. Atiyah and I.M. Singer. The index of elliptic operators III. *Annals of Mathematics*, 87:546–604, 1968.
11. M. Berger, P. Gauduchon, and E. Mazet. *Le spectre d'une variété Riemannienne*, volume 194 of *Lecture Notes in Mathematics*. Springer, 1971.
12. N. Berline, E. Getzler, and M. Vergne. *Heat kernels and Dirac operators*. Springer Verlag, New York, 1992.
13. S. Bochner. Curvature and Betti numbers. *Annals of Mathematics*, 49:379–390, 1948.
14. R. Bott. Lectures on Morse theory, old and new. *Bulletin of the American Mathematical Society*, 7:331–358, 1982.

15. R. Bott and L.W. Tu. *Differential Forms in Algebraic Topology*, volume 82 of *Graduate Texts in Mathematics*. Springer Verlag, New York, 1982.
16. R. Brooks. Constructing isospectral manifolds. *American Mathematical Monthly*, 95:823–839, 1988.
17. W. Browder. *Surgery on simply-connected manifolds*. Springer, 1972.
18. J. Cheeger, M. Gromov, and M. Taylor. Finite propagation speed, kernel estimates for functions of the Laplace operator, and the geometry of complete Riemannian manifolds. *Journal of Differential Geometry*, 17:15–54, 1982.
19. P.R. Chernoff. Essential self-adjointness of powers of generators of hyperbolic equations. *Journal of Functional Analysis*, 12:401–414, 1973.
20. A. Connes. A survey of foliations and operator algebras. In *Operator Algebras and Applications*, pages 521–628. American Mathematical Society, 1982. Proceedings of Symposia in Pure Mathematics 38.
21. A. Connes. Non-commutative differential geometry. *Publications Mathématique de l'Institut des Hautes Études Scientifiques*, 62:41–144, 1985.
22. A. Connes. *Non-Commutative Geometry*. Academic Press, 1995.
23. A. Connes and H. Moscovici. Cyclic cohomology, the Novikov conjecture, and hyperbolic groups. *Topology*, 29:345–388, 1990.
24. G. de Rham. *Differentiable manifolds*. Springer, 1984.
25. J. Dieudonné. *Eléments d'analyse*. Gauthier-Villars, Paris.
26. J. Dodziuk. L^2 harmonic forms on complete manifolds. In S.T. Yau, editor, *Seminar on Differential Geometry*, pages 291–302. Princeton, 1982. Annals of Mathematics Studies 102.
27. S.K. Donaldson and P. Kronheimer. *The geometry of four-manifolds*. Oxford University Press, 1990.
28. H.M. Donnelley and F. Xavier. On the differential form spectrum of negatively curved Riemannian manifolds. *American Journal of Mathematics*, 106:169–185, 1984.
29. N.T. Dunford and J.T. Schwartz. *Linear Operators Part II: Spectral Theory*. Wiley, 1963.
30. M. Freedman and F. Quinn. *Topology of 4-manifolds*, volume 39 of *Princeton Mathematical Series*. Princeton University Press, 1990.
31. E. Getzler. Pseudodifferential operators on supermanifolds and the Atiyah-Singer index theorem. *Communications on Mathematical Physics*, 92:163–178, 1983.
32. E. Getzler. A short proof of the local Atiyah-Singer index theorem. *Topology*, 25:111–117, 1986.
33. P.B. Gilkey. Curvature and the eigenvalues of the Laplacian for elliptic complexes. *Advances in Mathematics*, 10:344–381, 1973.
34. P.B. Gilkey. *Invariance Theory, the Heat Equation, and the Atiyah-Singer Index Theorem*. Publish or Perish, Wilmington, Delaware, 1984.

35. C. Gordon, D.L. Webb, and S. Wolpert. One cannot hear the shape of a drum. *Bulletin of the American Mathematical Society*, 27:134–138, 1992.
36. P. Griffiths and J. Harris. *Principles of Algebraic Geometry*. Wiley, New York, 1978.
37. M. Gromov. Kähler hyperbolicity and L^2 Hodge theory. *Journal of Differential Geometry*, 33:263–292, 1991.
38. M. Gromov and H.B. Lawson. Positive scalar curvature and the Dirac operator. *Publications Mathématiques de l'Institut des Hautes Études Scientifiques*, 58:83–196, 1983.
39. B. Helffer and J. Sjostrand. Puits multiples en mécanique semi-classique. IV. Etude du complexe de Witten. *Communications in PDE*, 10:245–340, 1985.
40. F. Hirzebruch. *Topological Methods in Algebraic Geometry*. Springer, 1978, 1995.
41. W.V.D. Hodge. *Harmonic Integrals*. Cambridge, 1941.
42. M. Kac. Can one hear the shape of a drum? *American Mathematical Monthly*, 73:1–23, 1966.
43. G.G. Kasparov. Equivariant KK-theory and the Novikov conjecture. *Inventiones Mathematicae*, 91:147–201, 1988.
44. S. Kobayashi and M. Nomizu. *Foundations of differential geometry*. Wiley-Interscience, 1963 and 1969.
45. T. Kotake. An analytical proof of the classical Riemann-Roch theorem. In *Global Analysis*, volume 16 of *Proceedings of Symposia in Pure Mathematics*, pages 137–146. American Mathematical Society, 1970.
46. S. Lang. *Algebra*. Addison-Wesley, 1995. Third edition.
47. H.B. Lawson and M.L. Michelsohn. *Spin Geometry*. Princeton, 1990.
48. A. Lichnerowicz. Spineurs harmoniques. *Comptes Rendus de l'Académie des Sciences de Paris*, 257:7–9, 1963.
49. J. Lohkamp. Metrics of negative Ricci curvature. *Annals of Mathematics*, 140:655–683, 1994.
50. W. Lück. L^2 invariants of regular coverings of compact manifolds and CW complexes. In R.J. Daverman and R.B. Sher, editors, *Handbook of Geometry*. Elsevier, 1998.
51. H.P. McKean and I.M. Singer. Curvature and the eigenvalues of the Laplacian. *Journal of Differential Geometry*, 1:43–69, 1967.
52. R.B. Melrose. *The Atiyah-Patodi-Singer Index Theorem*, volume 4 of *Research Notes in Mathematics*. A.K. Peters, Wellesley, Massachusetts, 1993.
53. J.W. Milnor. On manifolds homeomorphic to the seven-sphere. *Annals of Mathematics*, 64:399–405, 1956.
54. J.W. Milnor. *Morse Theory*, volume 51 of *Annals of Mathematics Studies*. Princeton, 1963.

55. J.W. Milnor and J.D. Stasheff. *Characteristic Classes*, volume 76 of *Annals of Mathematics Studies*. Princeton, 1974.
56. S. Minakshishundaram and A. Pleijel. Some properties of the eigenfunctions of the Laplace operator on Riemannian manifolds. *Canadian Journal of Mathematics*, 1:242–256, 1949.
57. R. Palais. *Seminar on the Atiyah-Singer Index Theorem*, volume 57 of *Annals of Mathematics Studies*. Princeton, 1965.
58. V.K. Patodi. An analytic proof of the Riemann-Roch-Hirzebruch theorem for Kähler manifolds. *Journal of Differential Geometry*, 5:233–249, 1971.
59. R. Plymen. Strong Morita equivalence, spinors, and symplectic spinors. *Journal of Operator Theory*, 16:305–324, 1986.
60. D. Quillen. Superconnections and the chern character. *Topology*, 24:89–95, 1985.
61. J. Roe. An index theorem on open manifolds I. *Journal of Differential Geometry*, 27:87–113, 1988.
62. J. Roe. An index theorem on open manifolds II. *Journal of Differential Geometry*, 27:115–136, 1988.
63. J. Roe. *Index theory, coarse geometry, and the topology of manifolds*, volume 90 of *CBMS Conference Proceedings*. American Mathematical Society, 1996.
64. W. Rudin. *Functional Analysis*. McGraw-Hill, 1973.
65. L. I. Schiff. *Quantum Mechanics*. McGraw-Hill, 1968. Third edition.
66. W. Schmid. Representations of semismiple Lie groups. In M.F. Atiyah, editor. *Representation theory of Lie groups*, pages 185–235. Cambridge University Press 1979.
67. R.T. Seeley. Elliptic singular integral equations. In *Singular Integrals*, volume 10 of *Symposia in Pure Mathematics*, pages 308–315. American Mathematical Society, 1967.
68. B. Simon. *Trace ideals and their applications*. Cambridge University Press, 1979.
69. E. Spanier. *Algebraic Topology*. McGraw-Hill, 1966.
70. S. Stolz. Positive scalar curvature metrics: existence and classification questions In *Proceedings of the International Congress of Mathematicians, Zürich 1994* volume I, 2, pages 625–636, Basel, 1995. Birkhauser.
71. M. Taylor. *Pseudodifferential Operators*. Princeton, 1982.
72. C.T.C. Wall. *Surgery on Compact Manifolds*. Academic Press, 1970.
73. E. Witten. Supersymmetry and Morse theory. *Journal of Differential Geometry* 17:661–692, 1982.
74. J.M. Ziman. *Elements of advanced quantum theory*. Cambridge, 1969.

Index

algebra of equivariant operators, 194
ansatz, 123
asymptotic expansion, 101
    of heat kernel, 161
Atiyah, M.F., collected works of, 148
Atiyah-Singer index theorem, 164
    historical discussion of, 146

Betti number, 91, 183
Bianchi identity, 14, 32
bundle
    frame, 23
    principal, 23

characteristic class, 30
Chern character, 35
    relative, 66
Chern class, 33
Chern-Weil theory, 30
Christoffel symbol, 10, 11, 29
Clifford algebra, 41
Clifford bimodule, 50
Clifford bundle, 43
    graded, 43, 141
    Riemann endomorphism of, 47, 156
Clifford contraction, 43
complex manifold, 51
complex manifold, 175
connection, 9, 24
    induced, 25
    Levi-Civita, 12
curvature, 10, 28
    Ricci, 14, 48
    Riemann, 13, 155
    scalar, 15, 49, 170
    twisting, 48, 64
cyclic cohomology, 150
cylindrical end, 177

de Rham
    cohomology, 87
    operator, 51
Dirac complex, 87
discrete series, 201
divergence, 20, 45
Duhamel's principle, 98

elliptic estimate, 77
eta function, 178
Euler characteristic, 140
Euler class, 39

filtered algebra, 151
finite propagation speed, 104
fixed point, 135
    simple, 136
formal power series, 158
four-dimensional geometry, 175, 181
Fourier series, 71
framing, 29
    synchronous, 29
functional calculus, 83, 127

Garding inequality, 186
Garding's inequality, 76
genus
    $\widehat{A}$, 36, 169
        of complex manifold, 176
    $\mathcal{L}$, 36, 173
    Chern, 34
    Pontrjagin, 35
geodesic, 15
    coordinates, 16, 99, 160
geometric endomorphism, 133
Getzler filtration, 154
Getzler symbol, 156, 158
    constant part of, 159
    well defined, 161
graded algebra, 151
    associated, 152

grading, 141
    canonical, 142
graph, 78
Green's operator, 88

half-spin representation, 62
harmonic oscillator, 119
heat equation, 95
    harmonic oscillator, 123
heat kernel, 96, 157
    approximate, 97
    asymptotic expansion of, 99
Hessian, 126
higher index theory, 170
Hodge
    star operation, 19, 172
    theorem, 88
homomorphism-like property, 152
Hopf conjecture, 201
horizontal, 25
hyperbolic space, 131, 201

index, 143
    multiplicativity of, 147, 193
interior product, 49
intersection, 91, 172

Laplacian, 19
    spectrum of, 115
Lefschetz number, 133
localization, 106

McKean-Singer formula, 145, 199
Mehler's formula, 124, 163
mollifier, 79, 84
Morse function, 185
Morse inequalities, 184, 191
Morse lemma, 189
Morse theory, 183, 192

operator
    annihilation, 120
    creation, 120
    de Rham, 51

Dirac, 43
    with coefficients, 51
Dolbeault, 52
Hilbert-Schmidt, 110, 145
    polynomial coefficient, 155
signature, 172
smoothing, 79, 113, 193
trace class, 111
unbounded, 80

Pfaffian, 33, 39
Pin group, 57
Poincaré duality, 89
polynomial
    Hermite, 121
    invariant, 30
Pontrjagin class, 33
Pontrjagin genus, 64

quaternions, 53, 171

rapidly decreasing function, 83
representation
    spin, 51
rescaling, 166

Schwartz space, 106
Sobolev space, 72, 74
spin bundle, 63
Spin group, 57
spin representation, 61
spin structure, 63
$Spin^c$ group, 67
$Spin^c$ structure, 68
superalgebra, 55
superbundle, 38
superconnection, 39
supersymmetry, 135
supertrace, 141, 143
symbol map, 152

tensor, 13
    algebra, 152
    antisymmetric, 17

theorem
>   Atiyah Γ-index, 200
>   Atiyah-Bott-Lefschetz, 137
>   Atiyah-Patodi-Singer, 178
>   Atiyah-Singer, 164
>   Bochner, 45, 91
>   Gauss-Bonnet, 148, 180
>   Hirzebruch signature, 174
>   Hodge, 88
>   Hopf, 140
>   Hopf-Rinow, 16
>   Karamata, 116
>   Lichnerowicz, 170
>   Lidskii, 111
>   Parseval, 72
>   relative index, 179
>   Rellich, 73, 117
>   Riemann-Roch, 149, 176
>   Rochlin, 174
>   Sobolev embedding, 73, 109, 195
>   spectral, 81
>   Stokes, 19

trace, 111, 141, 197
>   relative, 62, 142

transposition, 58

vertical, 23
volume form, 18, 56

wave equation, 95, 128
Weitzenbock formula, 44, 48, 157
Witten complex, 125, 185

zeta function, 117, 178